Jochen Joos

Microstructural Characterisation, Modelling and Simulation of Solid Oxide Fuel Cell Cathodes

Schriften des Instituts für Angewandte Materialien –
Werkstoffe der Elektrotechnik
Karlsruher Institut für Technologie

Band 30

Eine Übersicht aller bisher in dieser Schriftenreihe
erschienenen Bände finden Sie am Ende des Buchs.

Microstructural Characterisation, Modelling and Simulation of Solid Oxide Fuel Cell Cathodes

by
Jochen Joos

KIT Scientific Publishing

Dissertation, Karlsruher Institut für Technologie
KIT-Fakultät für Elektrotechnik und Informationstechnik

Tag der mündlichen Prüfung: 8. Dezember 2015
Hauptferentin: Prof. Dr.-Ing. Ellen Ivers-Tiffée
Korreferentin: Prof. Dr. Dagmar Gerthsen
Korreferent: Dr. Thomas Carraro

Impressum

KIT Scientific Publishing

Karlsruher Institut für Technologie (KIT)
KIT Scientific Publishing
Straße am Forum 2
D-76131 Karlsruhe

www.ksp.kit.edu

Print on Demand 2017 – Gedruckt auf FSC-zertifiziertem Papier

ISSN 2365-8029
ISBN 978-3-7315-0625-6
DOI: 10.5445/KSP/1000064791

Microstructural Characterisation, Modelling and Simulation of Solid Oxide Fuel Cell Cathodes

Zur Erlangung des akademischen Grades eines

DOKTOR-INGENIEURS

von der Fakultät für

Elektrotechnik und Informationstechnik
des Karlsruher Instituts für Technologie (KIT)

genehmigte

DISSERTATION

von

Dipl.-Math. techn. Jochen Joos
geb. in Konstanz

Tag der mündlichen Prüfung:	8.12.2015
Hauptreferentin:	Prof. Dr.-Ing. Ellen Ivers-Tiffée
Korreferentin:	Prof. Dr. Dagmar Gerthsen
Korreferent:	Dr. Thomas Carraro

Acknowledgements

First and foremost, I would like to express my deepest gratitude to my supervisor, Prof. Dr.-Ing. Ellen Ivers-Tiffée, for her valuable guidance, encouragement and excellent advices. I could not have wished for a more supportive mentor and I am deeply grateful for the trust she has placed in me, for providing an excellent working environment and the possibility to gain further experience by various conferences and research trips. Her constant promotion of my professional and personal development is not taken for granted.

I would also like to thank Prof. Dr. Dagmar Gerthsen from the Laboratory for Electron Microscopy (LEM) at KIT for her interest in my work, the pleasant and fruitful discussions with her and her group members regarding all aspects of electron microscopy and for her acceptance to act as one of the secondary reviewers for my thesis.

Just as well I would like to thank Dr. Thomas Carraro from the Institute for Applied Mathematics at the Heidelberg University for being my mentor regarding all matters of simulation and numerical mathematics and the excellent years of collaboration. Without him this thesis would never be what it is. His relationship as advisor, colleague, and friend is one that I cherish.

The members of my dissertation committee, Prof. Dr. rer. nat. Michael Siegel, Prof. Dr. rer. nat. Friedrich Jondral and Prof. Dr.-Ing. Wolfgang Heering, I want to thank for their time and support. The Friedrich-und-Elisabeth-BOYSEN-Stiftung and the Deutsche Forschungsgemeinschaft (DFG) are gratefully acknowledged for funding, as well as the Steinbuch Centre for Computing (SCC) for giving me access to their computing facilities.

Prof. Koichi Eguchi and his group are deeply acknowledged for being kind enough to welcome me for three month at the Kyoto University, thus enabling me to gather invaluable and wonderful experience.

During the time of developing this thesis, I was fortuned to work with a great group of people at my institute, which I would like to thank: The foundation of this work has been laid by my diploma thesis advisor and colleague, Bernd Rüger, who has been a great teacher to me. I would also like to thank my group leader, André Weber for his constantly support and for sharing his expertise and knowledge. Furthermore, I would like to thank Wolfgang Menesklou for his guidance, especially with the hardware tools such as our SEM and for his friendly attitude.

All colleagues and teammates I want to deeply thank, not only for their support during the last years, but moreover for creating a pleasant working atmosphere and for all the great experience apart from our professional work. The group has grown too large to mention all of them, but in deputy I want to thank Moses Ender, Helge Geisler, Jan Hayd, Jörg Illig, Dino Klotz, Michael Kornely, Alexander Kromp, André Leonide and Julian Szasz for being more of friends than of colleagues. Christian Gabi is gratefully acknowledged for being available 24/7 to support me with all technical matters, as well as Torsten Johannsen and Stefan Ziegler together with the team of the mechanical workshop. I would like to thank Sarah van den Hazel for sample preparation and her support regarding all chemical aspects as well as Andrea Schäfer and Marika Schäfer for their constant help in all organizational matters.

Special thanks also go to my students Andreas Häffelin, Andreas Messner, Gina Steinke, Ingo Rotscholl, Anne Wannenwetsch, Christine Dörflinger and Florian Wankmüller, whose committed and motivated work has significantly contributed to this thesis.

Jacob Packham, Kimberly McLeod, Andreas Messner and Ivan Russel are gratefully acknowledged for proofreading and editing the English manuscript.

Finally my warmest, heartfelt thank goes out to my friends and my family for all their support during the process. My wife Stephanie and my wonderful kids Yumi and Milo had to alternatingly endure and renounce me in this partly exhausting time, but they were nevertheless always there for me and fully supported me in all matters – each of them in his own way. To my parents, my brother and my sister, I am grateful for your support. Especially without the support of my parents, this thesis would not exist. I wish my father would still be here to see it.

Jochen Joos

Karlsruhe, December 2016

Zusammenfassung

Die Verteuerung und Verknappung fossiler Brennstoffe sowie ein stetig steigender Energiebedarf forcieren die Bemühungen, neue Technologien zu entwickeln, die diese Energieressourcen effizienter nutzen. Brennstoffzellen, wie z.B. die Festelektrolyt-Brennstoffzelle SOFC (engl., Solid Oxide Fuel Cell), sind hierfür eine vielversprechende Technologie, da sie die chemische Energie von Brennstoffen direkt und sehr effizient in elektrische Energie umwandeln. Da hierbei keine Schadstoffemission auftritt, können Brennstoffzellen einen wesentlichen Beitrag zur effizienten und umweltschonenden Energiegewinnung leisten.

Gründe, weshalb sich die Brennstoffzelle trotz vieler Vorteile bisher nicht großflächig gegen bereits etablierte Technologien zur Stromerzeugung durchsetzen konnte, sind vor allem in zwei Nachteilen zu finden: Zum einen muss die Langzeitstabilität im Betrieb erhöht und zum anderen die Kosten deutlich reduziert werden. Hieraus ergibt sich der dringender Bedarf die verwendeten Komponenten, insbesondere die beiden Elektroden Anode und Kathode, zu verbessern. Die Leistungsfähigkeit der Elektroden hängt gleichermaßen von ihren Materialeigenschaften sowie der porösen Elektrodenstruktur ab. Deshalb ist es unerlässlich, ein umfangreiches Verständnis der in den Elektroden ablaufenden Prozesse sowie deren Zusammenhang mit der Mikrostruktur zu gewinnen. Die Bezeichnung *Mikrostruktur* wird verwendet, da die Größe der Partikel und Poren der Struktur im Bereich von Mikrometern und darunter liegt.

Um die Leistungsfähigkeit und Langzeitstabilität der Brennstoffzelle zu verbessern, lag der Fokus der Forschung lange Zeit auf der rein elektrochemischen Charakterisierung der Zelle, um die Identifikation, Trennung und Quantifizierung der verschiedenen Verlustprozesse zu ermöglichen. Einen vollständigen Zusammenhang zwischen den Verlustprozessen und der Mikrostruktur der Elektroden herzustellen war jedoch nicht möglich, da hierfür auch eine exakte mikrostrukturelle Charakterisierung notwendig ist. Etablierte Methoden wie die Analyse von Bruchflächen der Zelle im Rasterelektronenmikroskop (REM) reichen nicht aus, um die dreidimensionale (3D) Struktur umfassend zu analysieren. Dies wurde erst in den letzten Jahren durch Weiterentwicklungen im Bereich der 3D Rekonstruktionstechniken, insbesondere der FIB (focused ion beam) Tomographie, möglich.

Die vorliegende Arbeit beschäftigt sich mit der mikrostrukturellen Charakterisierung, Modellierung und Simulation poröser SOFC-Kathoden. Vorrangiges Ziel dieser Arbeit war es, Methoden zu Entwickeln, um das komplexe Zusammenspiel zwischen Elektrodenmikrostruktur, Materialeigenschaften, Langzeitstabilität und Leistungsfähigkeit zu verstehen und dadurch die Optimierung der Elektrodenmikrostruktur zu ermöglichen. Im Folgenden werden die maßgeblichen Ergebnisse dieser Arbeit kurz zusammengefasst.

3D Rekonstruktion von porösen SOFC-Elektroden

Grundvoraussetzung um das komplexe Zusammenspiel von Mikrostruktur und Leistungsfähigkeit der Elektrode verstehen zu können, ist die detaillierte Kenntnis der dreidimensionalen Mikrostruktur. Aus diesem Grund wurde im ersten Hauptkapitel die FIB Tomographie, bei der aus einer Sequenz von 2D Schnittbildern eine 3D Repräsentation der Probe erzeugt wird, adaptiert für die Rekonstruktion von SOFC-Elektroden. Um die in dieser Dissertation vorgestellten Ergebnisse mit denen anderer Forschungsgruppen vergleichen zu können, wurde in Abschnitt 4.1 zunächst ein Überblick über den Stand der weltweiten Entwicklungen in Hinblick auf SOFC-Elektrodenrekonstruktion gegeben.

Im Laufe dieser Arbeit konnten viele Verbesserungen bezüglich Datengewinnung und Bildbearbeitung erzielt werden, welche auch von anderen Gruppen übernommen wurden. Beispielsweise wurde die Infiltration des Porenraums der Elektrodenstruktur mit Epoxidharz optimiert. Dies führte zu signifikanten Verbesserungen in Bezug auf einen vollständig automatisierten Rekonstruktionsprozess, insbesondere bei der Bildverarbeitung (Abschnitt 4.2). Die herkömmliche Vorpräparation der Probe wurde verbessert (vergleiche Abb. 2.9 und 4.4), was Abschattungseffekte, Helligkeitsunterschiede zwischen aufeinanderfolgenden REM-Aufnahmen und Probleme durch Materialablagerungen minimierte. Eine Methode zur verbesserten Analyse von Multiphasenelektroden sowie Strukturen mit isolierten Poren wurde entwickelt, welche die Informationen von zwei Detektoren des Rasterelektronenmikroskops nutzt. Dabei wird der Kontrast zwischen den Phasen optimiert, indem jeder Detektor für die Identifikation verschiedener Phasen optimal eingestellt wird.

Methoden für die Verarbeitung der Bilddaten und der strukturellen Analyse wurden entwickelt und etabliert. Die eindeutige und automatische Identifikation der vorhandenen Phasen (*Segmentierung*) ist entscheidend für eine exakte Mikrostrukturanalyse großer Volumen, weshalb zwei Segmentierungsmethoden entwickelt und diskutiert wurden. Die erste Methode ist die häufig verwendete Schwellenwert-Segmentierung (engl. *Thresholding*), für die Methoden zur Identifikation des optimalen Schwellenwerts evaluiert wurden. Als zweite Methode wurde ein selbstentwickelter Algorithmus präsentiert, der zu den *Region-Growing*-Segmentierungsmethoden gehört. Dieser erlaubt eine automatisierte Segmentierung auch für Bilddaten mit schwachem Phasenkontrast. Obwohl dieser Algorithmus für die Segmentierung von Strukturen mit drei Phasen eingeführt wurde, kann er durch geringe Modifikationen auch auf zwei oder mehr Phasen angewendet werden.

Die akkurat segmentierten Daten repräsentieren die tatsächliche Elektrodenmikrostruktur und bilden die Grundlage für die quantitative Mikrostrukturcharakterisierung. Strukturparameter wurden berechnet, was einen Vergleich verschiedener Elektrodenstrukturen ermöglichte und beispielsweise Aussagen über betriebsbedingte Änderungen in der porösen Elektrodenstruktur erlaubte. Die Methoden und Algorithmen zur Berechnung von Volumenanteilen, Oberflächen, Tortuositäten und Größenverteilungen der einzelnen Phasen in den Elektroden wurden hierbei sorgfältig evaluiert. Sie stellen die konsequente Weiterentwicklung und Anwendung existierender Methoden dar.

Ein Hauptaugenmerk lag auf der Identifizierung und Beseitigung potentieller Fehlerquellen, die im Laufe des mehrstufigen Rekonstruktionsprozesses auftreten können. Insbesondere wurde eine detaillierte Fehleranalyse bezüglich der Auflösung sowie der benötigten Volumengröße im Hinblick auf die Repräsentativität der 3D-Bilddaten durchgeführt. Deren Einflüsse auf die Genauigkeit der Ergebnisse wurde analysiert und Richtlinien definiert (Abschnitt 4.8). Für den aktuellen Kathodentyp des Forschungszentrums Jülich (Typ 2; durchschnittliche Partikelgröße 552 nm) wurde festgestellt, dass ein Volumen, welches eine Größe von mindestens 15^3 Partikeln (ca. 580 μm^3) besitzt, eine ausreichend gute Statistik enthällt und somit als repräsentatives Volumenelement (RVE) angesehen werden kann. Außerdem stellt eine Auflösung von mindestens 10 Voxel pro Partikeldurchmesser eine untere Grenze für die Auflösung dar. Diese Erkenntnisse können in zukünftigen Untersuchungen verwendet werden, um die Einstellungen für eine exakte Rekonstruktion zu spezifizieren.

Unter Berücksichtigung all dieser Erkenntnisse war es möglich, verschiedene Kathodentypen mit hoher Auflösung (25 bis 35 nm Voxelgröße) und den größten Volumina, die bisher in der Literatur beschrieben wurden (bis zu 32 049 μm^3), zu rekonstruieren. Damit konnte z.B. erstmals gezeigt werden, dass sich die Mikrostruktur von LSCF Kathoden im Betrieb über mehr als 1000 h bei verschiedenen Betriebstemperaturen (600°, 750° und 900 °C) praktisch nicht verändert (Kathoden vom Typ 2 in Abschnitt 4.9). Unterschiede in der Mikrostruktur aufgrund unterschiedlicher Sintertemperaturen wurden ebenso quantifiziert (Kathoden vom Typ 3). Mit steigender Sintertemperatur nahmen Porositätsanteil, Oberflächendichte und Tortuosität des Materials ab, da sich Strukturen mit größeren Partikeln bildeten.

Modellierung und Simulation von mischleitenden Kathoden

Das zweite Hauptkapitel dieser Arbeit beschäftigt sich mit der Modellierung und Simulation von mischleitenden (engl., mixed ionic-electronic conducting, MIEC) Kathoden. Geeignete Leistungsmodelle sind notwendig, um die Einflüsse des Materials und der Mikrostruktur auf die Leistungsfähigkeit voneinander trennen zu können. Ein kurzer Überblick über bisher veröffentlichte Modelle wurde in Abschnitt 5.1 gegeben. Bei der Modellierung ist es wichtig, die Mikrostruktur so genau wie möglich zu berücksichtigen. Aus diesem Grund ist eine direkte Verwendung der rekonstruierten Mikrostrukturen für die Simulationen am besten. Ein ParCell3D genanntes 3D FEM Leistungsmodell wurde präsentiert, welches die rekonstruierten Mikrostrukturen als Berechnungsgrundlage zur Ermittlung des flächenspezifischen Kathodenwiderstands ASR_{cat} (engl., area-specific resistance, ASR) als Leistungsindex verwendet (Abschnitt 5.2). Dieses Modell wurde in Zusammenarbeit mit T. Carraro (Universität Heidelberg) entwickelt und stellt eine Weiterentwicklung des am Institut für Angewandte Materialien (IAM-WET) von B. Rüger entwickelten Modells dar (Abschnitt 5.1.3). Mit Hilfe dieses Modells konnte die erste Leistungssimulationen einer LSCF Kathode basierend auf detaillierten 3D-Tomographiedaten der komplexen Kathodenmikrostruktur durchgeführt werden. Die Berechnung des Kathodenwiderstands eines gegebenen Materialsystems in Abhängigkeit der tatsächlichen Mikrostruktur kann die unterschiedlichen Einflüsse auf die Leistungsfähigkeit getrennt sichtbar machen.

In ParCell3D – wie auch in anderen MIEC-Kathodenmodellen – ist das Material durch den chemischen Diffusionskoeffizienten D^δ und den chemischen Oberflächenaustauschkoeffizienten k^δ repräsentiert. Der Einfluss dieser Parameter auf den berechneten ASR_{cat} wurde untersucht (Abschnitt 5.3), wobei sich zeigte, dass die große Streuung der Literaturwerte dieser Parameter auch folgerichtig zu einer großen Streuung des berechneten ASR_{cat} führt. Deshalb wurde eine exakte Methode zur Bestimmung dieser Materialparameter an porösen MIEC-Kathoden vorgestellt (Abschnitt 5.4). Diese steht im Kontrast zu den meisten veröffentlichten Methoden, welchen dichte Sinterproben zugrunde legen. Die evaluierte Methode ermöglichte zum ersten Mal eine in-situ Beobachtung der Veränderungen von D^δ und k^δ im laufenden Betrieb. In Verbindung mit den Ergebnissen der mikrostrukturellen Quantifizierung vor und nach dem Betrieb konnte gezeigt werden, dass die Degradation fast ausschließlich durch das Material verursacht wird: Bei einer Betriebstemperatur von 600 °C war die signifikante Leistungsabnahme durch die Degradation von sowohl D^δ als auch k^δ verursacht. Bei 750 °C wurde eine geringere Degradation beobachtet als bei 600 °C, hauptsächlich verursacht durch eine nichtlineare Verschlechterung von D^δ um eine Dekade, während k^δ nahezu unverändert blieb. Nur geringe Degradation konnte bei 900 °C beobachtet werden, die aber erneut auf eine Verschlechterung der Sauerstoffionendiffusion D^δ im Material zurückgeführt werden konnte.

Die Materialparameter D^δ und k^δ wurden auch mit Messergebnissen des IAM-WET evaluiert (Werte von A. Leonide; Abschnitt 5.4.3) und zeigten gute Übereinstimmung mit den in der Vergangenheit berechneten Werten. Außerdem konnte gezeigt werden, dass Berechnungen des ASR_{cat} mit ParCell3D und diesen Werten, basierend auf der rekonstruierten Kathodenstruktur als Berechnungsgrundlage nahezu perfekt mit den gemessenen Werten übereinstimmten, was darauf schließen lässt, dass die verwendeten Parameter korrekt sind.

Eine detaillierte Analyse der volumenbezogenen Repräsentativität für die Berechnung des ASR_{cat} wurde ebenso durchgeführt. Es zeigte sich, dass ein RVE für diese Berechnung anisotrop sein kann. Zum einen muss das Volumen groß genug sein, um die mikrostrukturellen Parameter genau wiederzugeben, zum anderen muss die Höhe des betrachteten Volumens die gesamte aktive Dicke der Kathode beinhalten. Dabei zeigt das Volumen über der aktiven Dicke nahezu keinen Einfluss, solange der Sauerstofftransport in den Poren und der Elektronentransport im Material nicht gehemmt sind. Wieviel der Kathode tatsächlich elektrochemisch aktiv ist (beginnend am Interface Elektrolyt/Kathode), hängt von den Werten für D^{δ} und k^{δ} und der Mikrostruktur ab. Abhängig von der aktiven Dicke konnte die betrachtete Grundfläche kleiner (im Falle einer großen aktiven Dicke) oder größer sein (im Falle einer geringen aktiven Dicke; siehe Abschnitt 5.5.1).

Die Einflüsse der Mikrostruktur auf die Leistungsfähigkeit der Kathode wurden ebenso untersucht. Es wurde gezeigt, dass von der besten bis hin zur schlechtesten Mikrostruktur (hergestellt durch unterschiedliche Sintertemperaturen), der berechnete ASR_{cat} um etwa den Faktor zwei zunimmt. Jedoch zeigte sich durch Vergleiche der Simulationsergebnisse mit den Messwerten, dass für eine Untersuchung des reinen Mikrostruktureinflusses auf die Kathodenleistungsfähigkeit (also unter Ausschluss anderer Einflüsse, wie z.B. Zweitphasenbildung am Elektrolyt/Kathoden-Interface) zunächst Untersuchungen am Modell nötig sind. Aus diesem Grund wurde ein Modell entwickelt, welches realistische computergenerierte Mikrostrukturen erzeugt (Abschnitt 5.7). Dieses stochastische Mikrostrukturmodell simuliert den Sinterprozess und es konnte gezeigt werden, dass dabei synthetische Mikrostrukturen erzeugt wurden, die in ihren Charakteristika nahezu identisch zu den rekonstruierten Mikrostrukturen sind. Dies wurde durchgeführt, um das stochastische Mikrostrukturmodell zu parametrieren und zu validieren.

Basierend auf diesem stochastischen Modell wurden realitätsnahe Mikrostrukturen mit unterschiedlichen Charakteristiken generiert. Anschließend wurde von diesen Strukturen der flächenspezifische Widerstand berechnet und für die Leistungsfähigkeit günstige Mikrostrukturcharakteristika identifiziert. Es wurde gezeigt, dass sich der ASR_{cat} in etwa halbiert, wenn die durchschnittliche Partikelgröße um den Faktor fünf reduziert wird. Mit der Erzeugung von Mikrostrukturen unterschiedlicher Porosität konnte gezeigt werden, dass für eine gute Verbindung des Porennetzwerks (d.h. vernachlässigbare Anteile von isolierten Poren und damit auch isolierter Oberfläche) eine minimale Porosität von 20 % nicht unterschritten werden darf. Die Oberflächendichte war für Porositäten zwischen etwa 40 und 60 % am höchsten. Die optimale Porosität ergab sich zu etwa 35 %.

Viele praktische Anwendungen sind für die vorgestellten Methoden denkbar. Zum Beispiel kann der Herstellungsprozess optimiert werden, indem man die Auswirkungen unterschiedlicher Herstellungsparameter auf die Mikrostruktur analysiert. Ergebnisse der mikrostrukturellen Quantifizierung können als Kontrollinstrument bei der großtechnischen Herstellung mit einer etablierten Fabrikationsmethode dienen. Außerdem ist nun die quantitative Analyse von mikrostrukturellen Veränderungen in den Elektroden unter verschiedenen Betriebsbedingungen möglich und wird letztendlich dazu beitragen, das Zusammenspiel von Elektrodenmikrostruktur, Material, Langzeitstabilität und Leistungsfähigkeit besser zu verstehen.

Das Erreichen der in dieser Dissertation gesteckten Ziele ermöglicht nun die modellbasierte Optimierung der Kathodenmikrostruktur.

Contents

1. Introduction

The rising cost and scarcity of fossil fuels, as well as the rising levels of environmental pollution resulting from their extraction and combustion, drive the development of new and efficient technologies for the production and storage of electrical energy. One such promising technology is the fuel cell. It directly converts chemical energy into electrical energy at a very high efficiency. The losses from thermal and mechanical energy conversion in fuel cells are significantly lower than for conventional combustion systems, resulting in reduced fuel consumption and CO_2 emissions. Moreover, when operated in reverse mode, fuel cells can quickly and efficiently store energy as hydrogen, produced by high-temperature electrolysis. These are, naturally, clear environmental benefits.

Of the several classes of fuel cell, Solid Oxide Fuel Cells (SOFCs) have attracted huge interest as a bridge technology towards a sustainable energy future. Despite their clear advantages, the large-scale use of SOFCs for energy production and storage has yet to be achieved. There are two reasons for this: first, their long-term operation stability must be improved and, secondly, their cost must be reduced. These problems point to the need to further improve and optimize the components and materials used in SOFC single cells. The electrochemical performance of single cells mainly depends on their material composition and the microstructure of the porous electrodes. It is therefore essential to improve the understanding of the processes taking place in electrodes and their relationship to the microscopic electrode structure. The term *microstructure* is used because the structural features are in the micrometers range or smaller.

Huge efforts in SOFC research and development have been made to electrochemically characterise fuel cells. For instance, the identification, separation and quantification of different SOFC component losses have been studied, e.g. by applying electrochemical impedance spectroscopy (EIS) together with adequate analysis techniques. However, a detailed characterisation of electrode microstructure is crucial and represents the missing link between, on the one hand, occurring processes and losses and, on the other, electrode performance. This is mainly because common microstructural analysis methods (such as scanning electron microscopy (SEM)) only provide information in two dimensions, so the true nature of the three-dimensional (3D) electrode microstructure cannot be accurately assessed.

The situation has changed rapidly in the last few years with improvements to 3D reconstruction techniques. In particular, the development of focused ion beam (FIB) tomography has enabled the accurate 3D reconstruction of complex microstructures. In 2006 the first SOFC electrode reconstruction was published, which demonstrated the potential of this technique. It allows the quantification and comparison of different electrode microstructures and the observation of structural changes, both during manufacture and operation. Such studies are indispensable for the structural optimization required to maximize SOFC performance and improve electrode durability. For the first time it is now possible to obtain all relevant information on the complex relationships between electrode microstructure, material composition, long-term stability and performance.

Goals of this Thesis

The aim of this work is to understand the influence of microstructure on SOFC electrode performance, in order to optimize the microstructure. Therefore it is essential to obtain a detailed and reliable 3D characterisation of the electrode microstructure. Thus, the first goal of this work will be to adapt FIB tomography to the reconstruction of SOFC electrode structures. This will include all steps of the reconstruction process; from sample preparation and data acquisition up to the identification and separation of different phases within the acquired tomographic data.

After reconstruction the microstructure must be quantified to allow structural comparisons, observation of structural changes during operation or to link microstructure and performance. The second goal of this work is, therefore, to develop and adapt appropriate methods for quantifying the electrode microstructure. The results would be the values of microstructural parameters with which one can characterise the structure, e.g. volume fractions, particle sizes, surface areas and tortuosities.

The third goal of this work is to identify possible sources of error which may be introduced during the reconstruction process, from sample preparation to the final parameter value estimation. The insights gained here will be enormously useful when identifying the limitations of this technique and assessing results. Therefore, an essential part of this study is to provide resolution- and minimal volume guidelines. These must be considered during reconstruction, in order to determine the parameters to a very high degree of accuracy.

To predict SOFC electrode performance in relation to microstructure, all important microstructural features must be included in an electrochemical performance model. The simplest way to achieve this is by using homogenized models which can calculate, for example, the area specific resistance (*ASR*) of an electrode as a performance index using some microstructural parameters. Naturally, the microstructural data must be considered as accurately as possible. Accordingly, it is best to enable simulations which use the reconstructed microstructures directly. Hence the fourth goal of this work is to develop a performance model for the air-electrode (cathode). This should be able to directly use the reconstructed 3D microstructure as computational domain for calculating the electrochemical processes within the cathode. The model ought to predict the cathode resistance in dependency to its actual microstructure, and thus isolate the influence of material and microstructure on performance. This will provide understanding of the underlying processes.

When optimizing electrode structure, it is necessary to examine many electrodes of differing microstructure. However, due to the complex electrode manufacturing process, it is very complicated to design "real" microstructures that have exactly the features desired for investigation. Moreover, the FIB tomography reconstruction process is labour-intensive and time consuming. Therefore, alternative approaches are needed. The fifth and final goal of this work is to develop a numerical tool capable of generating artificial microstructures with realistic characteristics. These artificial structures should be usable as model geometry in the 3D performance models in exactly the same way as reconstructed "real" microstructures.

All the processes, methods and tools described above can be combined to achieve the overall aim of this work, the optimization of electrode microstructure.

Although most of the work presented here is on mixed ionic and electronic conducting (MIEC) cathodes, the methods and findings can also be applied for multiphase electrodes (e.g. Ni/YSZ anodes), which was also performed within this dissertation project. Therefore, characteristic features for reconstructing Ni/YSZ anodes will be intermittently mentioned at appropriate points in this thesis. More details can be found in the references, corresponding publications and the Appendix (Appendix A).

Outline

Chapter 2 lays the groundwork for this thesis by introducing its fundamentals. Following this, the cells and samples investigated in this work are introduced in Chapter 3.

The results are presented in the two main chapters, 4 and 5. Chapter 4 deals with the 3D reconstruction of SOFC electrodes via FIB tomography. First global activities in this field are summarized in Section 4.1, with a description of their current state. All steps of the reconstruction process via FIB tomography are presented and discussed in Sections 4.2 through 4.5, including e.g. sample preparation, data acquisition and image processing. In Section 4.6 adequate segmentation methods (e.g., assignment of each pixel to a specific phase of the electrodes) are described. The methods and algorithms developed for the quantitative structural analysis are presented in detail in Section 4.7, which includes accuracy aspects of the presented methods. A more detailed discussion of accuracy aspects regarding FIB tomography is given in Section 4.8. In the last section (4.9) of this chapter results from the individual sample quantification are presented. These constitute the basis for comparing the different electrodes.

Chapter 5 deals with the simulation of mixed conducting cathodes. It starts with an overview of existing models for the electrochemical simulation of MIEC cathodes (Section 5.1). In

Section 5.2 a 3D performance model is presented, capable of using the reconstructed 3D data directly as model geometry. The influence of material, represented in the performance models by the two parameters (1) surface exchange coefficient k and (2) solid-state diffusion coefficient D, is shown in Section 5.3. Different values for these parameters published in literature will be used for the simulations and the differences discussed. A method for the adequate determination of the said material parameters is presented in Section 5.4. This method enables one to separate the influences of material and microstructure on performance and degradation. Accuracy aspects of calculating the cathode resistance will be discussed in Section 5.5. The influence of microstructural parameters on performance is explained in Section 5.6 using a homogenized model from literature and the 3D performance model presented in Section 5.2. Section 5.7 introduces a numerical tool which mimics the sintering process for generating realistic but synthetic 3D microstructures. It is parametrized with the microstructural parameters derived from "real" cathodes via FIB tomography (as shown in Section 4.9). The chapter ends by presenting and discussing simulation results based on synthetic microstructures. These synthetic microstructures are generated with different microstructural characteristics, which enable the identification of advantageous microstructural characteristics and thus support the optimization of the cathode structure.

Finally, Chapter 6 briefly summaries the essential findings of this work.

2. Fundamentals

This chapter outlines the fundamentals upon which this thesis is built. The highly interdisciplinary character of this work is reflected by its different topics, which will be introduced in the following sections of this chapter. These sections are deliberately kept short, the interested reader is referred to the citations given on relevant text passages.

The first four sections deal with Solid Oxide Fuel Cells (SOFCs) and introduces their working principles (Section 2.1), loss mechanisms and current/voltage characteristics (Section 2.2), materials and their requirements if used in single cells (Section 2.3) and the importance of microstructure (Section 2.4). This provides a brief introduction to the fundamentals of SOFCs and their electrochemistry. It is mostly based on Refs. [1–3]. A deeper insight into these topics can be found e.g. in Refs. [1] and [4].

Afterwards, Section 2.5 provides an introduction to scanning electron microscope and FIB tomography, since these are the main methods used in this work. The chapter ends with a brief introduction to image processing (Section 2.6).

2.1. Working Principle of Solid Oxide Fuel Cells (SOFC)

An SOFC, as one of several different classes of fuel cells, is an electrochemical energy conversion device which produces electricity (and waste heat) by combining a gaseous fuel with an oxidant [1,4]. All reactants and products are continuously supplied and removed. Like all fuel cell classes, an SOFC consists of three main layers: an electrolyte between two electrodes (the oxygen electrode, or cathode, and the fuel electrode, or anode). The electrolyte in an SOFC is a dense, gas-tight but oxygen ion (O^{2-}) conducting ceramic membrane, which separates the fuel (e.g., hydrogen, H_2) and the oxidant (e.g., oxygen, O_2), thus preventing a direct combustion of the reactants:

$$\frac{1}{2}\mathrm{O_2}(g) + \mathrm{H_2}(g) \rightleftharpoons \mathrm{H_2O}(g). \tag{2.1}$$

Figure 2.1 shows the simplified functional principle and schematic structure of a solid oxide fuel cell. The overall redox-reaction (Eq. (2.1)) proceeds in two electrochemical partial reactions in the two porous electrodes, which are spatially separated by the dense electrolyte.

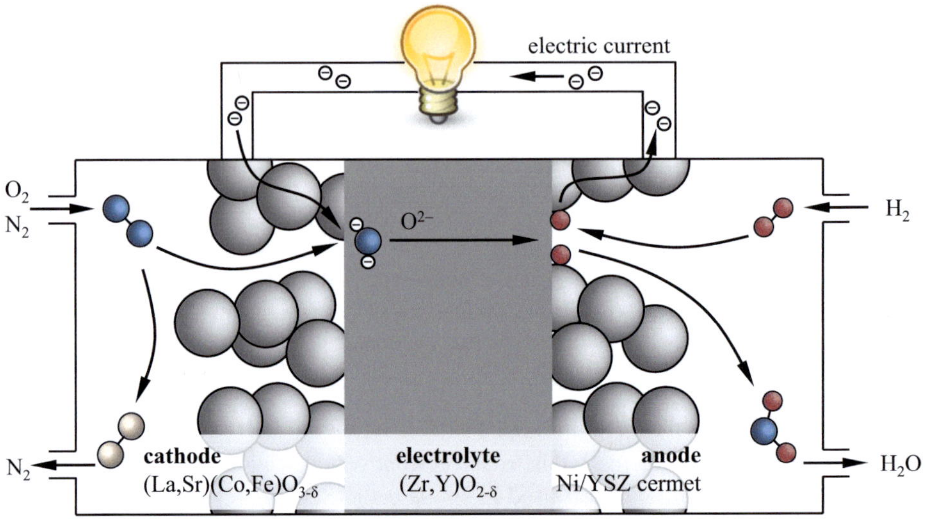

Figure 2.1.: Diagram of an operating SOFC (adapted from Ref. [5]), consisting of two porous electrodes (anode and cathode) separated by a dense electrolyte. The reduction and oxidation reactions are spatially separated, which enables the utilization of electrons involved in the redox process. Fuel and oxidant (here H_2 and O_2) are continuously supplied to the anode and cathode, respectively.

At the cathode, oxygen is reduced to oxygen ions (O^{2-}) by accepting two electrons from the cathode material, which positively charges the cathode:

$$O_{2,\mathrm{cat}} + 4e^-_{\mathrm{cat}} \rightleftharpoons 2O^{2-}_{\mathrm{el}}. \tag{2.2}$$

The generated oxygen ions are incorporated into the electrolyte phase, where they can diffuse towards the anode side. This ion diffusion j_{diff} through the electrolyte is induced by the large difference in chemical potential between the cathode (high oxygen partial pressure pO_2 of $0.01 \ldots 1$ atm) and the anode (low oxygen partial pressure $pO_2 = 10^{-27} \ldots 10^{-13}$ atm).

At the anode, the O^{2-} ions emerging from the electrolyte react with the H_2 to form water (H_2O). Thereby two electrons are released to the anode material, which negatively charges the anode:

$$2O^{2-}_{\mathrm{el}} + 2H_{2,\mathrm{an}} \rightleftharpoons 2H_2O_{\mathrm{an}} + 4e^-_{\mathrm{an}}. \tag{2.3}$$

Now, the released electrons are pulsed back by the negatively charged anode towards the positively charged cathode. As the electrolyte is not electronically conductive, the electrons are forced to flow back to the cathode through an outer circuit with an external load, thus performing electrical work. It should also be noted that the reaction can be forced into the reverse direction by applying a voltage to the device, thus electrolysing water.

If the external circuit is open, the electrons can not be transported back from the anode to the cathode. Hence an electric potential difference (denoted as Nernst voltage) introduces an electrical field in the opposite direction to the oxygen ion diffusion introduced by the chemical potential difference. After some time an electrochemical equilibrium will be reached, where the electric potential and the chemical potential differences are of equal magnitude, and the diffusion of oxygen ions will be impeded. The Nernst voltage U_{N} (or in this case also called open circuit voltage (OCV)) can be predicted using the oxygen partial pressures pO_2 of the anode and cathode

$$U_{\mathrm{N}} = \frac{RT}{n_e F} \ln \frac{\sqrt{pO_{2,\mathrm{cat}}}}{\sqrt{pO_{2,\mathrm{an}}}}, \tag{2.4}$$

where R is the universal gas constant, T is the absolute temperature, F is Faraday's constant and n_e the total number of electrons transferred ($n_e = 2$ in case of SOFC).

From a thermodynamic viewpoint, the Nernst voltage can also be calculated using the Gibbs free energy ΔG [6]

$$U_{\mathrm{N}} = -\frac{\Delta G}{n_e F}. \tag{2.5}$$

The Gibbs free energy is the energy available for conversion into usable work. It differs from the total energy by only the amount of energy lost as entropy. Taking the dependency of Gibbs free energy on temperature and oxygen partial pressure into account, U_{N} can be calculated by [1]

$$U_{\mathrm{N}} = -\frac{\Delta G_0(T)}{n_e F} - \frac{RT}{n_e F} \ln \left(\frac{pH_2O_{,\mathrm{an}}}{\sqrt{pO_{2,\mathrm{cat}}} \cdot pH_{2,\mathrm{an}}} \right). \tag{2.6}$$

In the typical SOFC operation range ($T = 600 \ldots 950\,^{\circ}\mathrm{C}$), when operated with hydrogen (with 1 % H_2O) as fuel and air as an oxidant, Eq. (2.6) yields voltages between 1.18 and 1.13 V [2]. Please note, these values apply to the theoretical Nernst voltage, without taking any losses into account. In this case, the theoretical Nernst voltage can also be denoted as U_{th}. However, in real cells all electrochemical processes have losses. Thus the cell voltage U_{cell} of a real system will always sit lower the theoretical Nernst voltage U_{th}. This will be discussed in the next sections.

2.2. Loss Mechanisms and Current/Voltage Characteristics of a Single Cell

During operation, where current is drawn from the system, irreversible processes occur which drop the cell voltage U_{cell} below the thermodynamically predicted voltage U_{th}. These voltage losses, called polarization or overpotential, originate in various phenomena that occur within the cell. The primary polarizations are: (a) ohmic, (b) concentration and (c) activation [7]. As more current is drawn from a cell, the losses increase non-linearly as qualitatively shown in Fig. 2.2. These losses and their origins will be discussed in the following sections.

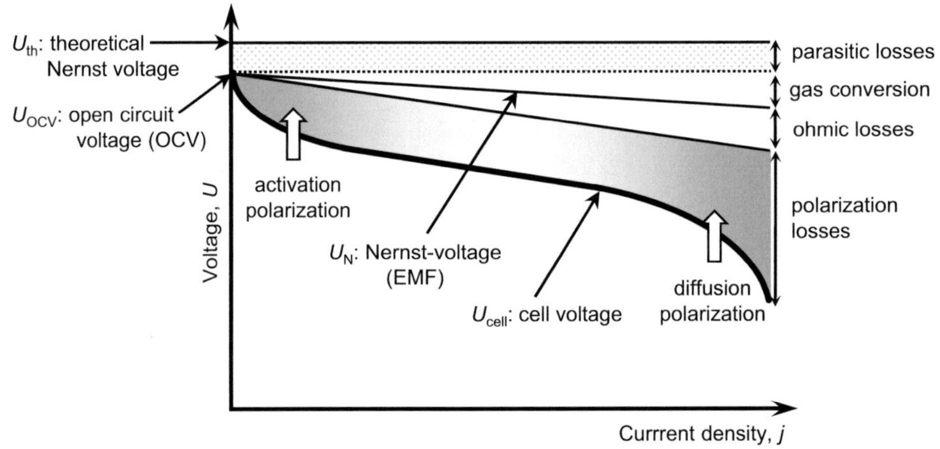

Figure 2.2.: Schematic plot of the C/V characteristics of an SOFC showing the different types of polarization (adapted from Refs. [7, 8]).

Parasitic Losses or Overpotential

As explained in the previous section, the open circuit voltage U_{OCV} is the driving force behind the reactions within the cell. In a real system, the OCV is smaller than the theoretically predicted Nernst voltage U_{th} due to different parasitic losses. Reasons for these overpotentials are e.g. undesired electron leaks across the electrolyte or even an imperfect gas-tight cell design or electrolyte. These reasons already reduce the electrochemical potential between anode and cathode at open circuit, thus lowering the Nernst-voltage.

Gas Conversion

Gas conversion losses due to fuel utilization also cause a drop in cell voltage. With increasing current density, the consumption of fuel and oxidant gas increases. Thus, the partial pressures of H_2 at the anode and of O_2 at the cathode decrease, while the partial pressure of the products increases with the increasing current density. At the anode, the partial pressure of oxygen

$pO_{2,\mathrm{an}}$ in the fuel gas mixture increases. As a consequence, U_N decreases with increasing current density, following Eq. (2.4). As already discussed, U_N represents the driving force for the overall cell reaction, and is therefore also called the electromotive force (EMF) [2].

Ohmic Losses

Ohmic losses arise from electronic and ionic transport through the electrodes and the electrolyte. The overall ohmic resistance R_ohm comprises the individual losses R_k. Obeying Ohm's law, the ohmic overpotential η_ohm is directly proportional to the current density j:

$$\eta_\mathrm{ohm} = j \cdot \sum_k R_k = j \cdot R_\mathrm{ohm}. \tag{2.7}$$

Since the conductivity of the electrolyte is orders of magnitude less than those of the electrodes, most of the ohmic losses in planar cells are due to the electrolyte, which is why thin electrolytes are ideal. In state of the art *anode-supported* fuel cells (see [9] for more information), the electrolyte thickness is around 10 µm or even thinner. The conductivity of 8YSZ (the most common electrolyte material, see Section 2.3) is $\sigma = 5\,\mathrm{S\,m^{-1}}$ at 800 °C [2, 10].

Activation Loss

Activation losses arise from electrochemical reactions that are required to transfer electrons to or from the electrode. During this stage a portion of the generated voltage will be lost as activation energy. Activation losses mainly occur at the so-called triple-phase boundary (TPB), where the ionic conducting-, electronic conducting- and gas phases meet. The relationship between activation overpotential and current density can be described using the Butler-Volmer equation [11]

$$j = j_{0,\mathrm{el}} \left[\exp\left(\alpha_\mathrm{el} \frac{n_\mathrm{e} F}{RT} \eta_\mathrm{act,el} \right) - \exp\left(-(1-\alpha_\mathrm{el}) \frac{n_\mathrm{e} F}{RT} \eta_\mathrm{act,el} \right) \right], \tag{2.8}$$

where $j_{0,\mathrm{el}}$ is the exchange current density of anode or cathode (which is temperature and partial pressure dependent), α_el is the apparent transfer coefficient and $\eta_\mathrm{act,el}$ is the activation overpotential of the according electrode. Thereby, α_el is an indicator of the symmetry of the activation energy barrier when a positive or negative potential is applied [2].

The activation losses can be reduced by lowering the required activation energy, which can be done by increasing the temperature, providing a catalyst or by supplying more reaction zones. Activation losses dominate at low current densities since the activation energy must be supplied to overcome the energy barrier that prevents spontaneous reactions [12].

Diffusion Overpotential Loss

Diffusion overpotential loss (or concentration polarization) is dominant at high current densities. It is caused by gaseous transport limitations in the porous electrodes, insufficient reactants at the TPBs, or inadequate product removal. This is generally a function of binary diffusion (that of O_2 and N_2 in the cathode and H_2 and H_2O in the anode) and electrode microstructure (e.g. porosity fraction and tortuosity of pores), as will be discussed in the next subsection. Naturally, the diffusion overpotential loss increases as electrode thickness increases.

If one or more diffusion processes are too slow, a concentration gradient occurs, resulting in an overpotential. These overpotentials (cathode polarization overpotential $\eta_{conc,cat}$ and anode polarization overpotential $\eta_{\text{conc,an}}$) can be described using the Nernst-equation (Eq. (2.4)) [13]:

$$\eta_{\text{conc,cat}} = \frac{RT}{4F} \ln\left(\frac{pO_{2,\text{cat}}}{pO_{2,\text{cat}}^{TPB}}\right) \tag{2.9}$$

$$\eta_{\text{conc,an}} = \frac{RT}{2F} \ln\left(\frac{pH_2O_{\text{an}}^{TPB} \cdot pH_{2,\text{an}}}{pH_2O_{\text{an}} \cdot pH_{2,\text{an}}^{TPB}}\right) \tag{2.10}$$

Thereby, $pO_{2,\text{cat}}$, $pH_{2,\text{an}}$ and pH_2O_{an} (in atm) are the known partial pressures at the gas channels, while the unknown partial pressures at the triple-phase points are denoted with a "TPB". The relationship between these parameters can be described by using diffusion equations like Fick's Law, as described e.g. in [14] and [2].

Overall, one should note that different losses occur inside a single cell, which are responsible for the characteristic course of the current-voltage curve shown in Fig. 2.2: parasitic losses decrease the cell voltage even at OCV conditions below the thermodynamically predicted voltage, while ohmic losses (mainly due to the electrolyte) increase linearly with increasing current density. Activation polarization is dominant at low current densities, while diffusion polarization is dominant at high current densities. Activation polarization and diffusion polarization are both mainly attributed to the porous electrodes.

2.3. Materials and Requirements of SOFC Components

In the long history of SOFCs, extensive material research has been done to optimize the basic components of a membrane electrode assembly (MEA), e.g. depicted in Fig. 2.1 (i.e. electrolyte, cathode and anode). Several comprehensive overviews on this subject can be found in literature [9, 15–19]. A very brief summary is hereby provided, introducing the materials used in the cells analysed in this thesis.

Naturally, the functionality of a MEA is of particular importance for the efficient operation of a SOFC. The MEA should allow highly efficient electrode reactions (Eqs. (2.2) and (2.3)), as well as ensure fast reactant transport. Hence, the electrodes must have a high catalytic activity and high electronic and ionic conductivity, combined with sufficient reaction sites and good transport paths for the different species. The latter point emphasizes the necessity for good electrode microstructure, which will be discussed in the next section. On the other side, catalytic activity defines the requirements of good electrode materials. The same holds true for the electrolyte, as will be discussed in the next subsection. Moreover, the individual components must be stable at high temperatures, economically viable and suitable for large scale manufacture. However, each component can not be optimized individually. The whole system of materials should also possess compatible thermal expansion behaviour to prevent the cracking and delamination that can result from a mismatch in thermal expansion coefficients. As a further consideration, materials ought not be reactive with other cell components during manufacture or operation and be stable over long-term operation.

The many requirements of the individual cell components have promoted extensive material research over the last decades. The most prominent materials used in state of the art SOFC components will be briefly mentioned in the following subsections.

Electrolyte

The main function of the electrolyte is to separate anode and cathode. It must be gas-tight and conduct oxide ions from the cathode to the anode side. Hence, it should provide significant ionic conductivity (σ_{ion}) and negligible electrical conductivity (σ_{el}) at operating temperatures. A conventional choice of electrolyte material in SOFCs is stabilized zirconium oxide (ZrO_2), since it fits the requirements best. Its high ionic conductivity results from the substitution of ZrO^{4+}-cations by di- or tri-valent rare earth or alkaline earth cations (e.g. Y, Sc, Ca, Mg, Nb, Sm, Yb) [20], of which Yttrium (Y) and Scandium (Sc) are typically used. The oxygen vacancies formed inside the crystal lattice by this substitution lead to an improved ionic conductivity, which is based on the hopping mechanism [21]. The most common electrolyte material is ZrO_2 doped with 8 mol% Y_2O_3 (denoted as 8YSZ in the following), due to its stability and high ionic conductivity [9, 20] (e.g. at 800 °C, $\sigma_{\text{ion}} = 5\,\text{S}\,\text{m}^{-1}$ [2, 10]). The thermal expansion coefficient in the temperature range of 30...800 °C of 8YSZ is $\alpha_{8\text{YSZ}} = 10.5 \cdot 10^{-6}\,\text{K}^{-1}$ [3, 16].

Next to 8YSZ, also 3YSZ (3 mol% Y_2O_3 in ZrO_2) has drawn attention due to its higher mechanical stability (higher fracture strength, more elastic). But it is not suitable for intermediate or low temperature applications because of its lower ionic conductivity.

For this lower temperature range, other ceroxide-based materials are under investigation. The most common dopants are Gadolinium (GDC, sometimes also denoted as CGO), Samarium (SDC) and Yttrium (YDC). However, one detriment is that for low oxygen partial pressures (as it is the case on the anode side), they became electronically conductive, especially at temperatures above 600 °C [22]. Despite this, these materials are still of interest for some applications [3]. In particular GDC is widely used in combination with YSZ, as will be discussed in the next section. For more details on electrolyte materials the reader is referred to Refs. [23] and [3] along with the references cited there.

Cathode

The cathode reduces oxygen gas to oxide ions. The most common cathode materials are ABO_3-type perovskite manganites, cobaltates, and ferrates (A = La, Sr, Ca, Ba; B = Mn, Co, Fe). Depending on the configuration, these perovskites can exhibit both pure electronic and ionic conductivity. Further important material requirements are chemical stability (long-term in highly oxidizing atmospheres), compatible thermal expansion behaviour and high electrochemical performance (which is mainly determined by the composition of the perovskite). These aspects are discussed at great detail in various publications, including [24–26].

Among these perovskites, mixed ionic electronic conducting (MIEC) cathode materials are used in state of the art SOFCs. Examples include lanthanum strontium cobalt ferrite (LSCF) and lanthanum strontium cobalt (LSC). In the past, barium strontium cobalt ferrite (BSCF) was also considered as a possible candidate. Their main advantage is that the entire surface of MIEC cathodes can (in principle) act as TPB, as will be discussed in Section 2.4. However, the drawback is that the direct contact of these materials to zirconia based electrolytes (such as YSZ) causes the formation of an insulating $SrZrO_3$ interlayer [15, 27–29]. To avoid this, one must apply a GDC interlayer between LSCF and the electrolyte [30, 31].

Beside these MIEC cathodes, also in widespread use is the pure electronic conducting perovskite strontium-doped lanthanum manganite (LSM). This is typically used in the composition $La_{1-x}Sr_xMnO_3$, with $x = 0.15 \ldots 0.25$ [32] and often together with YSZ in composite systems [33, 34].

Within this thesis, LSCF cathodes with a stoichiometry of $La_{0.58}Sr_{0.4}Co_{0.2}Fe_{0.8}O_{3-\delta}$, as fabricated by Forschungszentrum Jülich (FZJ) are applied, which has been extensively studied in recent years [24, 30, 31, 35–39].

Anode

The main functionality of the anode is to support the hydrogen (and/or carbon monoxide) oxidation reactions. It oxidises hydrogen to produce water and electrons. Hence, the main material requirements for the anode are high catalytic activity, high electronic and ionic conductivities, compatible thermal expansion and chemical stability (over long time periods in a highly reducing atmosphere). In anode-supported cells (ASCs), the anode structure provides the mechanical support to the MEA structure. High performance ASCs normally consist of a double-layer anode, with (1) a 300 to 1500 μm thick, highly porous anode substrate which provides mechanical stability and the transport of fuel, exhaust gases and electrons, and (2), a 5 to 30 μm thin anode functional layer (AFL) - which provides the electrooxidation of the fuel at the TPB.

Due to the reducing conditions, metals are preferred as anode materials. Next to expensive noble metals, nickel (Ni) is most suited, due to its high electrical conductivity (e.g. $\sigma_{el} > 10^5\,\mathrm{S\,m^{-1}}$ at $T = 1000\,^{\circ}\mathrm{C}$). The main disadvantage of nickel (as with other metals) is its high thermal expansion coefficient of $\alpha_{\mathrm{Ni}} = 18 \cdot 10^{-6}\,\mathrm{K^{-1}}$. To overcome this and adjust the thermal expansion coefficient, electrolyte material is mixed with metal [40], resulting in an ceramic metal compound called a *cermet*. As an example, for a cermet consisting of 75 mol% Ni and 25 mol% 8YSZ, the thermal expansion coefficient is reduced to about $13.5 \cdot 10^{-6}\,\mathrm{K^{-1}}$ [41]. An additional advantage of cermet anodes is that the anode reaction zone (Eq. (2.3)) is extended. This will be discussed in the next section.

The most common anode material is Ni/YSZ cermet due to its excellent characteristics, but Ni/ScSZ and Ni/GDC can also be used. The main advantages of Ni/YSZ anodes are the high catalytic activity of nickel and their low costs. Moreover, their fabrication is relatively cheap and can be performed with conventional ceramic processing techniques like tape casting [42].

A disadvantage of Ni/YSZ anodes is the fast kinetics of Ni oxidation at typical SOFC operation temperatures [43, 44]. Nickel tends to oxidize at high steam partial pressures, which are common when the cell is operated with a high fuel utilization [45]. Next to the deactivation as a catalyst, this transition from Ni to NiO gives a substantial volume increase and, therefore, mechanical stress. At worst, this can result in fatal damage to the cell assembly. For more details on anode materials, the reviews available in literature [46–48] are recommended.

Beyond the MEA, the practical operation of SOFC systems requires many additional components, which are commonly referred to as the Balance of Plant (BOP). However, as this work focuses only on cell components of a MEA, the interested reader is referred to [49] for more details regarding these components. In the next subsection, a description is given of the processes taking place within the porous electrodes, clearly showing the importance of microstructure.

2.4. Importance of Microstructure

The structure of porous electrodes contributes significantly to the electrode performance. To minimize the losses of a given material system of the MEA and maximize its performance, optimization of the electrode microstructure is the best possibility. When looking at the electrode polarization losses, it is obvious the diffusion polarization directly depends on the structure and morphology of the pores. The activation polarization depends on the available reaction sites (e.g. surface area and/or TPB density, connectivity and tortuosity of the phases, etc.). Although the most important electrode microstructure requirements seem obvious (good, short transport paths for the different species, sufficient active reaction sites, etc.), there is no consensus as to what constitutes a good microstructure [50]. The different effects often conflict, and the requirements depend on the rate limiting step [12,51]. This complicates optimization.

The following two subsections discuss the oxygen reduction at the cathode side, and the fuel oxidation at the anode side. Thereby, different categories of electrode materials will be considered.

2.4.1. Oxygen Reduction at the Cathode

The oxygen reduction at the cathode is a highly complex and fascinating field of research. To date, no final consensus exists as to the exact nature of the rate limiting processes [32]. Some detailed articles are available on this topic (e.g. Refs. [7,32,52]), here only a short summary is given.

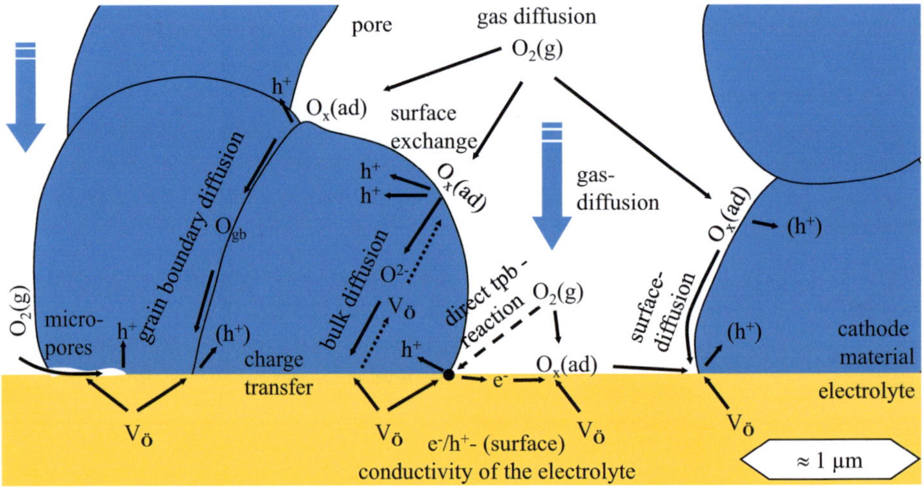

Figure 2.3.: Possible reaction pathways and involved species of the oxygen reduction reaction in a porous cathode structure (adsorbed oxygen species O_x(ad): adsorbed molecular oxygen $O_{2,ad}$, adsorbed oxygen atom O_{ad}, adsorbed oxygen ions O^-_{ad} and O^{2-}_{ad}, V^{oo}_O: oxygen vacancy, h^+: electron holes. Adapted from [32,53]).

Figure 2.3 shows possible pathways, including their elementary steps and species involved in the oxygen reduction reaction at the cathode. Factors determining the extent to which these reaction pathways contribute to the oxygen reduction include: the choice of cathode and electrolyte materials, the applied processing technique, the microstructural properties of the cathode layer, and the operating conditions [32]. For each pathway, one or more steps could be rate-determining steps (RDS). If only one pathway is rate-determining, the area specific resistance of the cathode, ASR_{cat} (a typically used indicator for the cathode performance) almost exclusively depends on this pathway. The possibility of a given pathway being active depends on the conductivity properties of the cathode material used. These materials can be divided into three categories: (a) pure electronic conducting materials (e.g. LSM), (b) two-phase composite materials consisting of an electronic and an ionic conducting material (e.g. LSM/YSZ) and (c) single-phase mixed electronic-ionic conducting materials (e.g. LSC, LSCF), cf. Fig. 2.4. The differences between these categories will be discussed in this section, and the parameters used to describe the pathways are introduced (according to Ref. [26]).

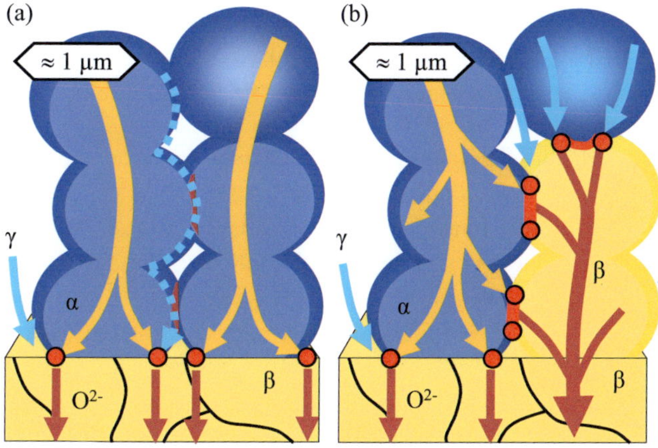

Figure 2.4.: Dominating reaction pathways and appropriate parameters for the different cathode categories: (a) pure electronic conducting, (b) two-phase composite, and (c) single-phase mixed electronic-ionic conducting (MIEC). Phases α, β and γ refer to electronic, ionic and pore phase, respectively (adapted from [26, 32]).

Pure electronic conducting LSM was widely used in early SOFC research. Only electrons can diffuse through LSM grains in the cathode layer, as schematically shown in Fig. 2.4a. Hence, the cathode reaction of Eq. (2.2) is restricted to (or at least near to) the TPB. In a pure electronic conducting cathode the TPB is constrained to the cathode/electrolyte interface (indicated by the red marks in Fig. 2.4a). Experiments on pattern model electrodes (strips of electrode material on an electrolyte) and oxygen isotope exchange experiments helped to confirm that the reactions occur within the vicinity of the TPBs [54,55]. It is worth noting, that Fleig *et al.*, [56] propose that an additional bulk path can open at high overpotentials due to the more reduced atmosphere.

Regardless, oxygen transport takes place either in the pores as gas diffusion or at the surface of the LSM grains as surface diffusion. Hence, the reaction pathway consists of parallel transport pathways for electrons in the cathode material and oxygen in the pores. The pathways are linked via charge transfer reaction at the TPB (described by the line-specific resistance LSR_{ct}). While conductivity values can be easily determined experimentally, values for LSR_{ct} are rarely available. By using pattern electrodes, the LSR_{ct} can be determined experimentally. Such experiments are difficult, but some values are still available [57,58]. In order to ensure enough oxygen at the TPB, the porosity must be high. But increasing porosity interrupts the electron conduction pathways. This again shows the necessity of microstructure optimization.

Pure electronic conducting cathodes show a large cathodic polarization resistance, especially when operated below 900 °C, due to the limiting effective TPB length [7]. To enhance the TPB length, dual-phase materials (composite materials) consisting of electronic conducting material (like LSM) and ionic conducting electrolyte material (like YSZ) can be used. As can be seen in Fig. 2.4b, pathways for the transport of oxygen ions also arise inside the cathode layer, which enables TPBs to also occur within the cathode volume. This can significantly lower the cathode polarization. Composite cathodes can be described with the same parameters used to explain pure electronic conducting cathodes on an ionic conducting electrolyte. Naturally, the existence of a third phase makes the structural optimization more complex, since three transport paths are now competing. In addition, the intimate contact of all three phases (electronic-, ionic- and pore-phase) at the TPB is required, the transport of the species involved in the reaction must be enabled. This includes that the ionic phase must be connected to the electrolyte, the ionic phase to the current collector and the pore phase to the gas channel. However, as with pure LSM cathodes, it is still a disadvantage that electrochemically active areas are limited to the TPBs.

For (single-phase) MIEC cathodes, the incorporation of oxygen occurs at the surface of the cathode particles rather than at the TPB lines, since the entire MIEC cathode surface can act as TPB. The potential reaction sites are therefore significantly increased (see Fig. 2.4c). For this reason, MIEC cathodes have the most potential for showing the lowest ASR_{cat}, indicating the best performance. One important goal in SOFC research is to lower the operation temperature, thus allowing cheaper alloys for housing and gas lines. But since most losses increase with decreasing temperature, huge efforts are made to use more sophisticated materials, especially at low temperatures. This makes MIEC cathodes like LSCF near indispensable when operating SOFCs at intermediate temperatures (IT-SOFC).

If a MIEC cathode with a high ionic conductivity is used, the reaction pathway via bulk diffusion is assumed to be dominant with a high degree of certainty [32, 59–62]. De

Souza [61] and Adler [60] suggest the following reaction path for oxygen at the MIEC surface and in the bulk: (i) dissociative oxygen adsorption, (ii) ionization of the oxygen atom, (iii) incorporation of the adsorbed oxygen ion in the cathode bulk, (iv) bulk diffusion, and (v) oxygen ion charge transfer from the cathode bulk into the electrolyte. It should be noted that it is not known which of these steps is rate determining, since there is no consensus in literature. The surface reaction steps (i), (ii) and (iii) are described by the chemical surface exchange coefficient k and step (iv) by the solid state oxygen ion diffusion coefficient D. These parameters are used for a quantitative description of the half-cell reaction kinetics (Eq. (2.2)) and depend on temperature and oxygen partial pressure (see Fig. 2.4).

According to [63,64], the material coefficients k and D can be categorized into three different types, which are normally assessed using dense bulk samples from chemical diffusion experiments (k^{δ}, D^{δ} or sometimes denoted as $\tilde{k}$, $\tilde{D}$), electrical conductance experiments (k^{Q}, D^{Q}) and tracer experiments (k^{*}, D^{*} or k^{tr}, D^{tr}). The different k and D values can be converted into each other using the thermodynamic factor $\gamma_o = \frac{1}{2} \cdot (\partial \ln p_{O2}) \cdot (\partial \ln c_o)^{-1}$ [65,66] and the equations $k^{Q} \approx k^{*}$ and $D^{Q} \approx D^{*}$, as well as $k^{\delta} = \gamma_o \cdot k^{*}$ and $D^{\delta} = \gamma_o \cdot D^{*}$ [26,60,63,64,66,67].

It should be noted that under some conditions also a competing route via surface diffusion of oxygen on the material can also be active, which has been observed for different materials (LSCF: [68], LSC: [69], LNO ($La_2NiO_{4+\delta}$: [70,71]).

2.4.2. Fuel Oxidation at the Anode

Ni/YSZ composite is by far the most widely used choice as anode material in SOFCs. Pure electronic conducting materials are impractically on the anode side, since there is no material that is simultaneously stable in the highly reducing atmospheres, matches the thermal expansion coefficient of the electrolyte and shows sufficient catalytic activity (cf. Section 2.3). Mixed conducting anode materials are also under investigation [47,72–74], but could never compete against composite anodes. Since in the framework of this thesis only technical relevant Ni/YSZ anodes were investigated, only this system will be discussed in the following.

The processes taking place on the anode during the oxidation reaction (Eq. (2.3)) are not fully understood and are subject to current research. Pattern and point model anodes are widely used to study the reaction kinetics as a function of T, pO_2 and pH_2O [75–79]. A good overview is given in [62]. This publication also presents possible reaction pathways for the oxidation of hydrogen at TPBs, as shown in Fig. 2.5. Although there is no consistent view on what the RDS is, Bessler *et al.*, [76] demonstrated that the hydrogen spillover is the active reaction pathway, as elementary kinetic models best fitted with experimental data for a wide range of conditions. Whatever the case may be, one common factor of all these pathways is that the reaction takes place within the vicinity of a TPB. Thus, the performance is expected to scale with the TPB length (l_{TPB}) according to [80, 81]

$$R_{\mathrm{pol}} \sim \frac{LSR_{\mathrm{ct}}}{l_{\mathrm{TPB}}}, \tag{2.11}$$

where R_{pol} is the electrode polarization resistance and LSR_{ct} the line-specific resistance. This naturally also holds for composite cathodes. As discussed for the cathode, determination of LSR_{ct} is difficult and values in literature for Ni/YSZ anodes vary by more than two orders of magnitude for the same indicated conditions. A summary of some values for Ni/YSZ can be found in [76] and [82], where possible reasons for this scatter are also discussed (e.g. inexact determination of TPB length, impurities at TPB, effects of pH_2O, etc.). As for cathodes, the connectivity of the phases must be ensured, in order for the TPBs to be active: the electronic conducting Ni must be connected to the current collector, the ionic conducting YSZ to the electrolyte, and the pore phase to the gas channel. When any of the three phases does not fulfil these requirements, the TPB will be inactive (or "isolated") and cannot contribute to the oxidizing reaction.

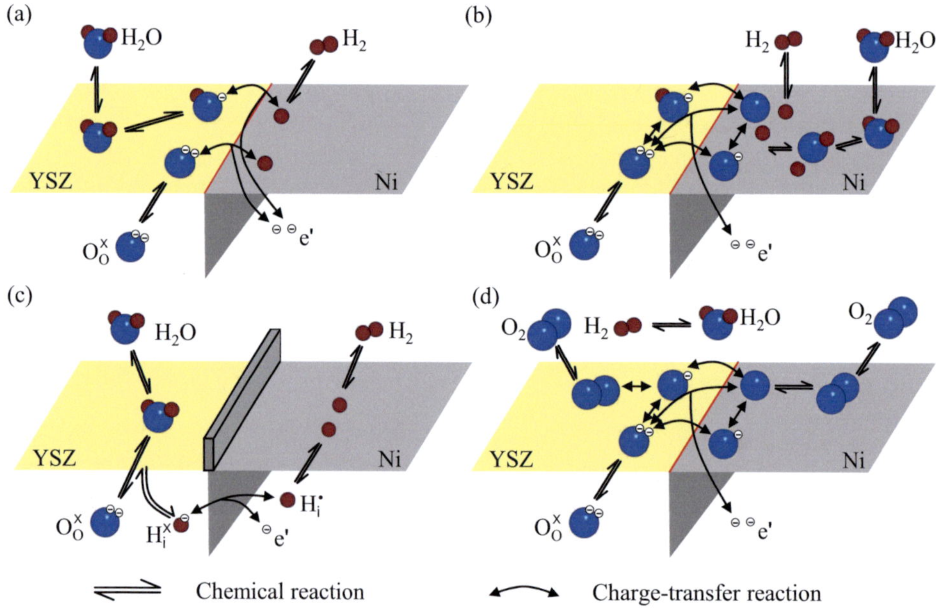

Figure 2.5.: Possible reaction pathways for the hydrogen oxidation at the TPB (a) hydrogen spillover, (b) oxygen spillover, (c) hydrogen interstitial and (d) oxygen evolution mechanism. The "wall" in panel (c) indicates potential segregated impurities (adapted from [76] [62]).

2.5. Scanning Electron Microscope (SEM) and Focused Ion Beam (FIB) Tomography

The following section provides an introduction on scanning electron microscopy (SEM) and focused ion beam (FIB) tomography, which are extensively used in this work. Since the microstructure is enormously important for the performance and long-term stability of the electrodes, methods for analyzing these structures are highly valuable. In the last decades, analysis of porous electrode microstructure has mainly been done using SEM, thus obtaining two-dimensional (2D) SEM micrographs of fractured or (machine-) polished cross-sections. Although this can provide useful microstructural information and enables the extraction of some basic microstructural parameters (like material fractions etc.), the true nature of the structure with its three-dimensional (3D) characteristics could not be assessed. A fundamental understanding of microstructural effects requires a 3D characterisation, as only this allows to obtain structural parameters of higher order topology such as tortuosity (a measure for the pathways of how convoluted they are) or the identification of bottlenecks. As a simple example, material fractions can be detected from 2D micrographs (as can surface area, albeit partly). Only a 3D analysis shows whether the phases are percolating (e.g. if LSCF is connected to both current collector and electrolyte, or the pores to the gas channel, see Section 4.7.1). If not, they can not be electrochemically active. Figure 2.6 shows a fractured (a) and a FIB polished (b) cross-section of a LSCF cathode with a GDC interlayer.

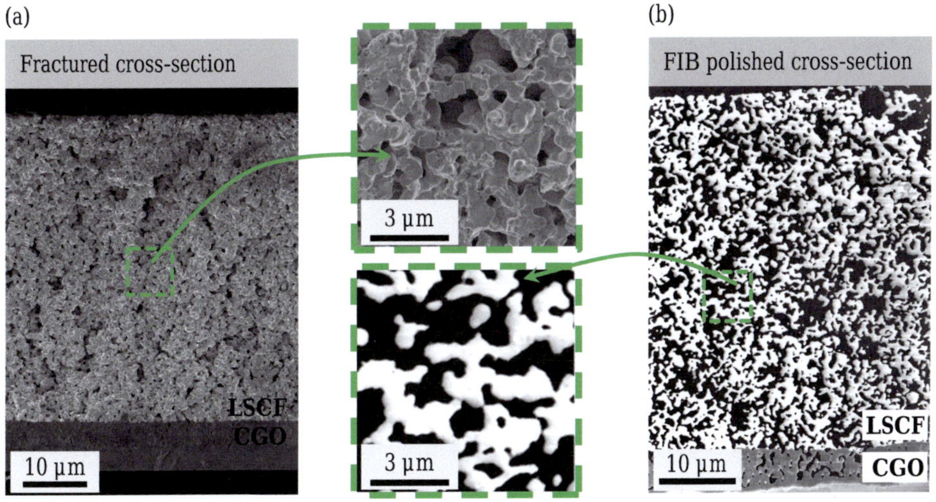

Figure 2.6.: SEM micrographs of a porous LSCF cathode and a GDC interlayer: (a) fractured and (b) FIB polished (pores are infiltrated with epoxy) cross-section [39].

The spatial structure of objects can be detected by tomographic methods. Two tomographic methods are commonly used to observe SOFC electrodes, FIB tomography and X-ray tomography [83, 84]. Considering the various pros and cons of the methods [85], FIB tomography is more suitable for the investigations done here on SOFC electrodes with particle sizes of 0.2 to 1 µm, as it allows for a higher resolution of down to ~10 nm compared to ~150 nm for lab-scale X-ray tomography (~20 nm performed at synchrotron).

FIB tomography is a relatively new method; the first reconstruction of an SOFC electrode was reported in 2006 [86]. Since then it has been widely used by many groups. A literature overview of reported reconstructions is provided in Section 4.1. For the first time FIB tomography enables the reconstruction and analysis of the spatial topology of SOFC electrodes. Thus, it is possible to get a more thorough understanding of the relations between microstructure and performance.

A FIB microstructure reconstruction is based on a stack of consecutive sectional images (cf. Fig. 2.10b), which are typically obtained by SEM. Naturally, the image quality is of enormous importance for the quality and reliability of the results of FIB tomography. Therefore, an introduction to SEM is given first, as the key to high-quality SEM images is a comprehensive understanding of the effects and the system parameters. It is mainly based on reviews [12, 87–90], where more detailed descriptions are to be found. Following this, FIB tomography will be introduced. More details on this can be found in [91–93]. Within this work, a Zeiss 1540 XB cross-beam, owned by the Institute for Applied Materials (IAM-WET) was used for the data acquisition. This system is equipped with an in-lens (IL) and an Everhart-Thornley detector (ETD, or also called SE2 detector).

2.5.1. Scanning Electron Microscope

In an electron microscope electrons are used for the imaging instead of light. The resolution of a microscope d_{res} (which in this content means the smallest distance between two points which can still be distinguished), is limited by the wavelength λ, e.g. according to the formula of Ernst Abbe

$$d_{\mathrm{res}} = 0.61 \frac{\lambda}{n_{\mathrm{b}} \sin \alpha}, \tag{2.12}$$

where n_b is the refractive index of the medium between the lens and the specimen, and α is half the angular aperture. The shortest wavelength of visible light (violet, $\lambda \approx 400\,\mathrm{nm}$) leads to a resolution limit of about 200 nm (see [94] for more information). Since electrons have a much smaller wavelength (3-12 pm), a much better resolution can be achieved with electron microscopes (at present about 0.1 nm). Other advantages over light microscopy are a higher depth of focus, the easy use of different contrast mechanisms for image creation and the simple interpretation of the images due to the 3D impression [89].

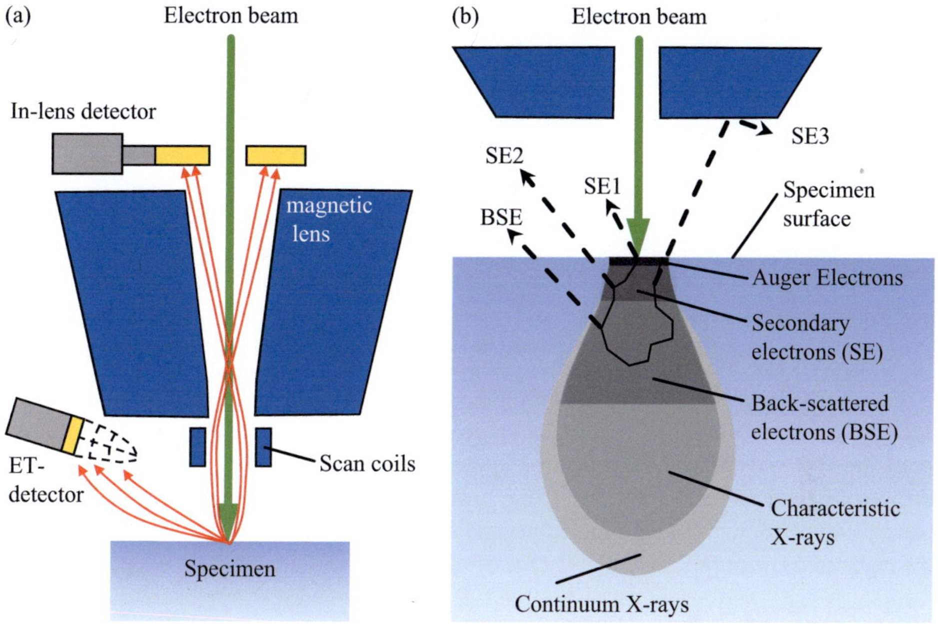

Figure 2.7.: (a) Schematic makeup of the beam column, from which the position of the detectors and possible electron trajectories can be seen. (b) The interaction between electron beam and specimen, showing the resulting interaction volume (or excitation bulb) and possibly generated signals (adapted from [89]).

A scanning electron microscope (SEM) is a special type of electron microscope [95]. In a SEM, a primary electron beam is created from an electron gun. This generates free electrons and accelerates them to energies between 1 and 40 keV. The beam is focused by magnetic lenses to create a small focused electron probe (smaller than 10 nm) on the surface of the specimen. The image is formed sequentially by scanning the focused electron beam across the specimen, which is how SEM got its name. Between each step the beam remains in a particular location for a fixed time (called dwell time), before it moves to the next position. Each location is represented as a pixel in the SEM image.

When the primary electron (PE) beam strikes the specimen, the electrons will scatter through the sample within a defined area called the interaction volume (or excitation bulb, see Fig. 2.7). Secondary products are created, which can be used for imaging and analysis: X-rays (which can be used to identify materials using an energy dispersive X-ray (EDX) system), heat and light are formed, but mostly secondary electrons (SE) and back-scattered electrons (BSE) are used to create the images.

Secondary electrons are low energy electrons (all emitted electrons with exit energies below 50 eV), which can only emerge from the specimen within a few nanometers from the sample surface (if produced deeper within the excitation bulb, they will simply be absorbed by the sample). The probability of emergence (correlated to the intensity of SE signal) decreases exponentially with distance from the surface. Emitted electrons can be classified as SE1, SE2 or SE3 signals, depending on whether they were generated by the beam (in the spot center) or at greater distance from the spot centre by back-scattered electrons, (cf. Fig. 2.7b or [96]). The SE are generated as ionization products of the PE beam, hence the signal tends to be higher for metals with lower ionization energy. However, in contrast to BSE there is no explicit relationship between secondary electrons yield and the atomic number of the elements [12].

Back-scattered electrons have a much higher energy (above 50 eV), which allows them to escape from deeper within the interaction volume. They are mainly generated by elastic collisions between the electron beam and the specimen atoms (see Fig. 2.7). Since the degree of back-scattering (denoted by the backscatter coefficient) strongly depends on the atomic number of the material, the information content of the BSE is mainly determined by the material contrast [89]. Meanwhile, the SE signal is more sensitive to the topology of the sample, as discussed below. However, the FIB/SEM system used here has an ETD as well as an in-lens detector. Thus, the images within this work are mainly based on SE, which provides sufficient contrast between the different phases when conducted at the low energy setting of the PE beam, as will be shown later.

The number of emitted (or more precisely *detected*) electrons produced at different locations provides the image contrast. If more electrons are detected at a particular location, a brighter spot appears on the image. Various factors influence the signal, especially with SE. For example, the surface properties significantly influence the obtained signal: more SE will be detected from raised areas and areas facing the detector (edge effect). On the other side, SE produced in a valley/hollow sometimes do not reach the detector (especially if the direct way to the detector is blocked by some parts of the sample) which leads to shadowing effects (cf. Section 4.2). Also, negatively charged areas appear brighter than positively charged areas and regions of heavy elements are brighter than those of lighter elements. Moreover, with increasing PE beam energy, the number of SE induced by primary electrons decreases, which means that the SE output drops. Different materials show a different characteristic of the scattering angles, which means that the SE signal as a function of the scattering angle depends on the material. The tilt angle of the specimen influences the exit angle and thereby the distribution of the SE and BSE. Increasing the incidence angle of the primary electrons can increase the SE output [87]. This is important for FIB tomography, where the observed sample surface is tilted by 36°, as explained in the next subsection.

A typical problem occurring during the acquisition of SEM images is noise. Image noise refers to a superposition of the image signal with a random fluctuation. Hence, the signal-to-noise ratio is a common measure of digital image quality. The intensity of noise in SEM images principally depends on the SEM adjustment parameters used for the image acquisition, e.g. acceleration voltage, probe current (aperture size) and working distance. SE3 type electrons contribute to noise since they are not emitted directly at the sample surface. Also the dwell time is of particular importance. As previously mentioned, the electron beam produces signals (e.g. SE and BSE) while hitting a specific point of the specimen. But the number of created interaction products varies, and therefore contributes to noise. Thus, increasing the dwell time per pixel reduces the noise level (or one can scan the sample several times and integrate the generated signal) [89].

It is often said that the backscatter coefficient for materials is independent of the acceleration voltage. But this is generally only true for voltages above 5 kV. Thyden *et al.* [96] showed with Monte Carlo calculations and experimentally, that the backscatter coefficient for Ni and YSZ clearly changes with acceleration voltage in the range of 1 – 5 kV. In their study the contrast increased with decreasing acceleration voltage. Another peculiarity of the sub-5 kV acceleration voltage range is that the backscattered coefficient still depends on the atomic number but it is not monotonously increasing with it. Thus, lighter elements can have a higher backscatter coefficient compared to heavier elements [87].

A very important parameter during SEM observation is the resolution of the SEM images. However, it is important to realize that details and resolution do not merely depend on the size of the electron probe and the distance between two scanned spot centres. The size and the characteristics of the excitation bulb are also important. The size of the excitation bulb is influenced by the acceleration voltage and the atomic number of the sample: the excitation bulb gets bigger for higher acceleration voltages and for samples with lower atomic number. Among other points (higher signal and better contrast between e.g. Ni and YSZ at lower voltages, less charging, etc.), this is one reason why relatively small acceleration voltages (e.g. 1.3 kV) are typically used during FIB tomography.

In order to only represent the sample surface (and not information from deeper within the sample), only the SE1 and SE2 type electrons should be used, as only they are created on or near the surface of the specimen. The SEM used within this work has two detectors: the Everhart-Thornley detector and the in-lens detector. A detector that mainly detects BSEs was not provided. An efficient way of detecting SE is by using the in-lens detector. It is located within the beam path of the primary electrons. Due to the viewing direction, images created with an in-lens detector seem to be flat, as they only contain relatively low topographic contrast. The major benefit of this detector is its high detection efficiency, particularly at very low acceleration voltages. It also almost exclusively detects SE [89]. For the samples investigated in this work the in-lens detector mostly showed a better contrast between the different materials (e.g. between nickel and YSZ) compared to the ETD. A disadvantage of this detector is its high sensitivity to sample charging: if electrons hit non-conducting or only partially conducting structures, these electrons accumulate at the surface, since they cannot discharge. Local charges are generated and affect the electron beam. This may significantly deteriorate the imaging quality (e.g. the sample seems to move and/or charged areas may seem very bright, etc.) [89].

Charging effects were found to be much less pronounced with the ETD. It is located on the side of the sample (cf. Fig. 2.7) and can also receive BSE, which enables a greater depth of field. Electrons moving to the detector are absorbed and directed by a collector. The collector voltage (suction voltage) can be varied (between −250 V and 400 V in 1 V steps; for standard application 300 V are recommended [89]), which influences the image characteristics. The collector voltage can be used to change the ratio between SE and BSE. Due to its position, images detected with ETD contain more topological information than in-lens images. While this is important for some applications, it should be noted that during FIB tomography the observed sample surface is flat due to FIB polishing. Hence, ideally, the consecutive images do not contain any topological information. A comparison of the two detectors can be seen in Fig. 2.8.

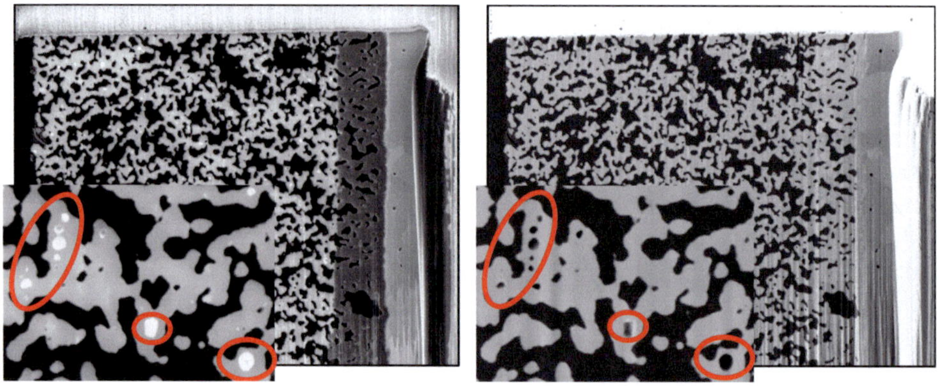

Figure 2.8.: FIB polished SEM images of a LSCF cathode with an GDC interlayer on a YSZ electrolyte, detected with an in-lens detector (left) and a ET detector (right). The in-lens image shows much stronger contrast between the LSCF and GDC phase, but is more sensitive to charging than the ETD image, as can be seen in the magnification. Contrast between LSCF and pores is equally good for both detectors.

In summary, the quality and resolution of electron micrographs depend on many different factors and parameters. Among many others, some very important factors are acceleration voltage (EHT), type of detector used to create the image, working distance, resolution, aperture size and dwell time of the electron beam. During FIB tomography, the working distance is fixed at the coincidence point of the electron and ion beam (at about 5 mm). Taking all these factors into account, it becomes obvious that creating the optimal SEM image can be highly complicated, depending on the nature of the specimen.

2.5.2. Focused Ion Beam Tomography

The functional principle of focused ion beam systems is in many respects similar to that of SEM systems. As the name implies, FIB is also based on a focused beam of charged particles (SEM: electrons; FIB: ions). In FIB systems, the ions are created from a source (which is also a field emitter) of liquid gallium (Ga). Both systems require a high vacuum environment and a beam rasters across the surface of the specimen to generate a signal that can be used for imaging [91]. But FIB can be used for more than imaging. For example, the ion beam can manipulate a specimen very precisely. A comprehensive discussion of the capabilities of FIB, as well as a broad description of the techniques, can be found in [91–93]. Here, only ion milling during FIB tomography is described.

A focused ion beam with a high beam current (typically between 100 pA to 20 nA for milling and 50 pA for imaging at 30 kV) is able to precisely remove (sputter) material from a specimen. By scanning the ion beam in a raster pattern, it is possible to obtain perfectly planar (FIB polished) cross-sections of a sample (cf. Fig. 2.6b). The beam current hitting the sample controls the image quality and the milling rate. It is controlled with apertures, located in the ion column [91]. The higher the beam current, the higher the removal rate. But lower beam currents tend to mill somewhat more precisely.

The milled plane is typically imaged with the electron beam of a SEM. Hence, state of the art systems are equipped with both ion and electron beams and the beams are tilted relatively to each other (e.g. by 54° in the Zeiss 1540 XB system used here; cf. Fig. 2.9a). Such combined focused ion beam / scanning electron microscope (FIB/SEM) systems are, therefore, called *crossbeam* (or *dual beam*) instruments. In contrast to single beam systems, the observed surface must not be tilted between milling and imaging.

Most reconstructions published in literature were done by applying the conventional bulk milling technique [12], where a cross-section is polished and a graduated trench is milled in the middle of the plane to expose the region of interest (ROI) prior to the sectioning (cf. Fig. 2.9). Thereby, the trench can be milled with relatively high beam currents of 5 to 20 nA, while the sectioning is typically done with lower beam currents between 100 pA to 2 nA. But this procedure can introduce various artefacts during image collection and has therefore been further developed in this work (see Section 4.3 and Fig. 4.4).

During sectioning, the ion beam is used to mill away slices of the sample with a defined thickness (this is the resolution in the slicing direction z, often between 20 and 50 nm). After milling, tilt-corrected SEM images are collected. Tilt correction is a necessary stage, since the image cross-section is not perpendicular to the SEM (cf. Fig. 2.9a). The top and bottom of the SEM image can be out of focus for the same reason. This can be avoided using the dynamic focus feature provided by the system. The sequence of FIB milling and SEM imaging repeats itself several hundred times to obtain several hundred of consecutive images.

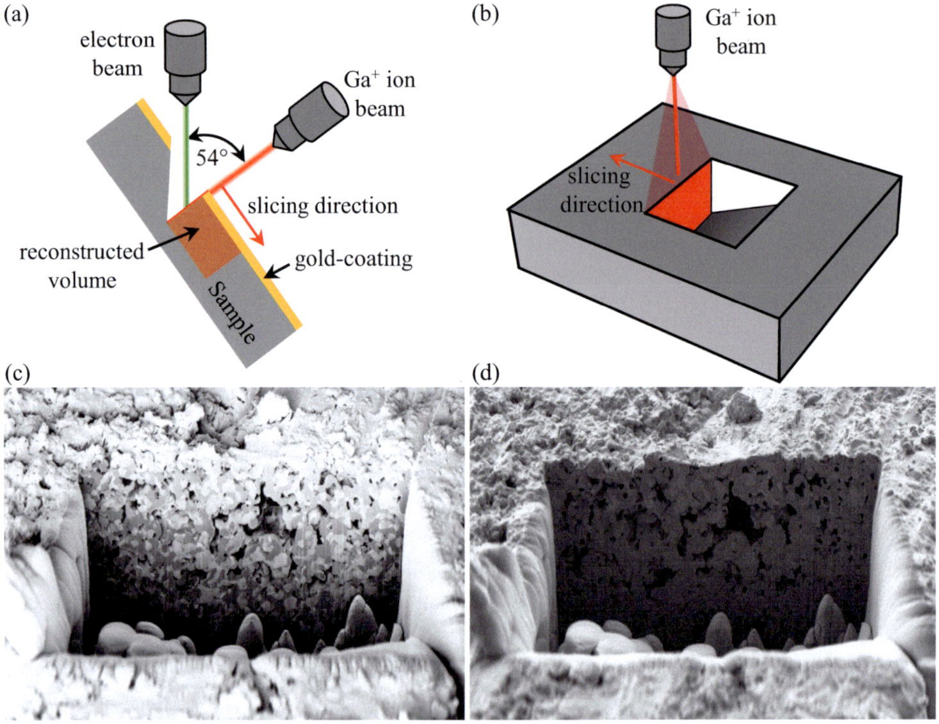

Figure 2.9.: (a) Schematic design of a FIB/SEM setup (adapted from [97]) and (b) conventional bulk milling technique, as applied by most groups (adapted from [12]), imaged with (c) in-lens detector and (d) ETD: a graduated trench has to be milled before the sectioning process can start to expose the region of interest (ROI). Note, that the conditions (e.g. phase contrast, pores not infiltrated) are not optimized for reconstruction.

Thereby, normally at least 15 µm of material are milled through in the slicing direction. A schematic showing the sample with the electron and ion beam assembly is shown in Fig. 2.9a. Figure 2.10a shows an example stack of consecutive SEM images.

The fully automated process of image collection can generally be done in two ways [12]: firstly, by sequentially milling and imaging, which means that the milling pauses whilst an image is captured and resumes after the imaging is done (often called *slice-and-view*). Second, by milling and imaging continuously, capturing images whilst the ion beam is still milling in the ROI. The latter method is much faster, but the two beams can interfere with each other, possibly resulting in image artifacts due to noise.

Both methods can be subject to common problems, various artefacts can occur during FIB/SEM serial sectioning. For example, some of the material which the ion beam sputters away can be re-deposited on the sample. This may affect the image quality, if it is re-

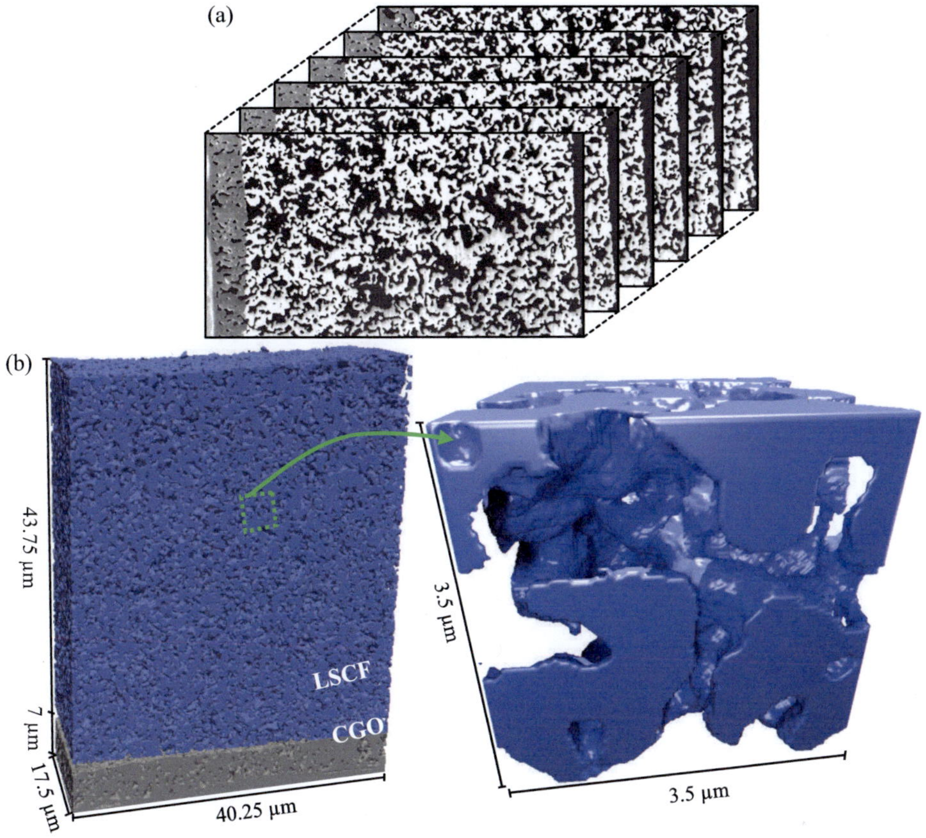

Figure 2.10.: (a) Example stack of consecutive SEM images, obtained by sequentially milling and imaging the sample surface. (b) 3D reconstruction from a part of the LSCF cathode and GDC interlayer (porosity is transparent), obtained via FIB tomography [39].

deposited on or near the exposed surface under investigation. Then the newly deposited layer is imaged instead of the actual sample structure. If a porous structure is investigated, re-deposition is even more problematic, as material can re-deposit inside the pores, thus influencing the yet to be examined microstructure. To avoid re-deposition inside the pores and to improve the stability of the porous structure, the pores can be backfilled with epoxy resin. This has the additional advantage that the cross sectional area is clearly defined (cf. Fig. 2.6b). It would be otherwise difficult to decide which regions of the image are in the cross-sectional plane, and which regions provide out-of-plane information from behind the current plane (cf. Fig. 2.9). It also supports the separation of the phases during image processing, as the epoxy within the pores appears almost black, which improves the contrast with the (grey) electrode materials. Altogether, impregnating the sample with epoxy resin helps to avoid artefacts in the FIB/SEM data.

Other undesired effects that often occur during sectioning can be seen in Fig. 2.9: the introduction of vertical streaking across the face of interest (*curtaining-effect* or *streaking*), charging inside the imaged plane, and a non-uniform contrast gradient across the face of interest (*shadowing*). Whether these undesired side effects appear (and to which degree) depends to a great extend on the preparation technique. Also the conditions during sectioning have a strong influence (e.g. the detector used for creating the image, cf. Fig. 2.9). Therefore, an advanced preparation procedure was developed within this work, which will be described in Chapter 4. This new procedure avoids these negative effects or at least minimizes them. This is a precondition for obtaining reliable results from FIB tomography.

Further key aspects have to be considered with respect to FIB/SEM data accuracy: first, the resolution which is necessary for the SEM images. Second, the size of the volume which has to be reconstructed. The resolution depends on the feature size under investigation (e.g. the particle sizes), since the smallest features of the analysed structure have to be sufficiently resolved. It is important that the reconstructed and analysed volume is large enough to be representative for the whole structure. How to set sufficient resolution and volume parameters is presented in Chapter 4.

Overall, FIB tomography is a suitable method for the 3D analysis of porous electrode microstructures. The main advantages are its high resolution (up to ~3 nm) and the high contrast between different materials, which enables the analysis of different phases in 3D structures. Additionally, one can collect the cross-sectional images with different detectors simultaneously (cf. Section 4.5). This allows combining of complementary information obtained by the different detectors.

A disadvantage of FIB tomography is that the method is destructive to the sample. Thus, it is not possible to perform further reconstructions from the same sample position, for example with other device parameters. Moreover, the analysed volume is limited and the sample preparation is labour intensive.

Further, after collecting the consecutive greyscale images, the stack of images has to be processed with appropriate methods. As a result, a 3D material distribution of the sample can be obtained (see Fig. 2.10b). The necessary steps to obtain an adequate reconstruction from the collected SEM images are introduced in the next section.

2.6. Introduction to Image Processing

Data sets obtained via FIB/SEM require image processing to improve the quality of 3D reconstructions. The aim of this section is to explain the fundamentals of image processing for SEM images, as far as described in Chapter 4. It is based on Refs. [98–100], where especially the latter reference is recommended for getting a more comprehensive introduction to this topic.

As mentioned in the previous section, the electrode microstructure is acquired as a stack of digital cross-sectional images. These digital images consist of discrete elements: the pixels. By stacking and aligning the FIB polished 2D images in a 3D space and expanding the image pixels in the axis of the slicing direction (space between two images), a 3D reconstruction of the structure consisting of *voxels* (volumetric pixels) is derived [101]. The 3D structure is represented as a $X \times Y \times Z$ matrix, where $X \times Y$ is the resolution of the SEM images and Z the number of images. Every voxel is an entry of this matrix, containing a discrete brightness value $g(x, y, z)$. In a typical 8-bit greyscale image, one can differentiate 256 discrete greyscale values from 0 (black) to 255 (white).

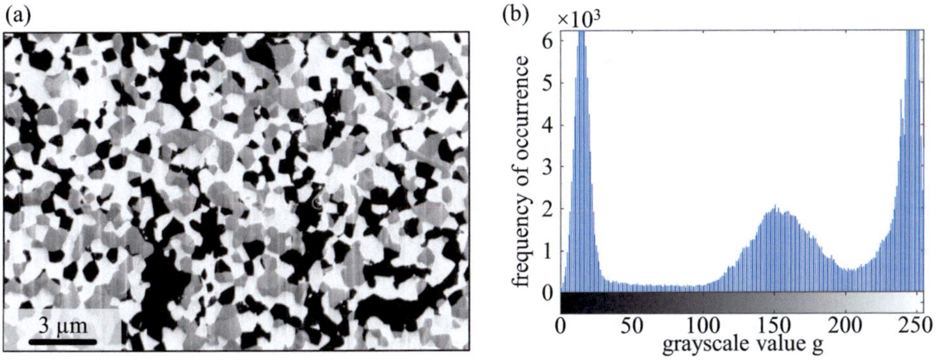

Figure 2.11.: (a) 2D cross-section of an Ni/YSZ anode as obtained by SEM with (b) corresponding histogram with the greyscale frequency distribution between 0 (black) and 255 (white). If the image quality is good, the peaks in the histogram are clearly separated, and the number of peaks corresponds to the number of phases in the structure (here: pores in black, Ni in grey and YSZ in white).

The voxels in such 3D structures are typically cubic. This is not absolutely necessary, but it greatly expedites further data processing. Cubic voxels result if the distance between the images corresponds to the pixel size of the SEM images. However, many groups have reported a difference between the resolution of the SEM images and the image distance [86, 97, 102, 103]. This can be readjusted by applying a re-sample step on the data, but this in itself can also lead to errors [104].

An important characteristic of 2D and 3D image-based data is the histogram. It plots the number of pixels or voxels $h(g)$ for each greyscale value g. A 2D cross-section of a porous Ni/YSZ anode and its corresponding histogram are shown in Fig. 2.11. An increased occurrence of a particular brightness value in the image results in a peak in the histogram. Hence if the image quality is sufficient, the number of peaks in the histogram corresponds to the number of phases inside the imaged structure. In an optimal case, the peaks (and hence the phases) are clearly separable from each other.

In reality, the image processing is complicated by image artefacts. This means imperfect or distorted digital image data, e.g. due to noise. As discussed in Section 2.5.1, noise is a common phenomena in SEM images, and the signal-to-noise ratio mainly depends on the adjustment parameters used to acquire the image. Figure 2.12 shows two SEM images at the same position of a porous LSCF cathode cross-section. The only difference between the images is the dwell time of the electron beam (or more precisely, the noise reduction option and the scan speed) for each pixel during image acquisition. The image in (a) was acquired by *line averaging* with a scan speed of five (image acquisition time: 40.1 s). Image (c) is from *pixel averaging* with a scan speed of six (image acquisition time: 2.6 s). As can be seen, the pixel averaging peaks are smaller but wider and tend to overlap with increasing noise level (Figs. 2.12b and d). It is therefore harder to determine the separation of the peaks, and ergo the phases.

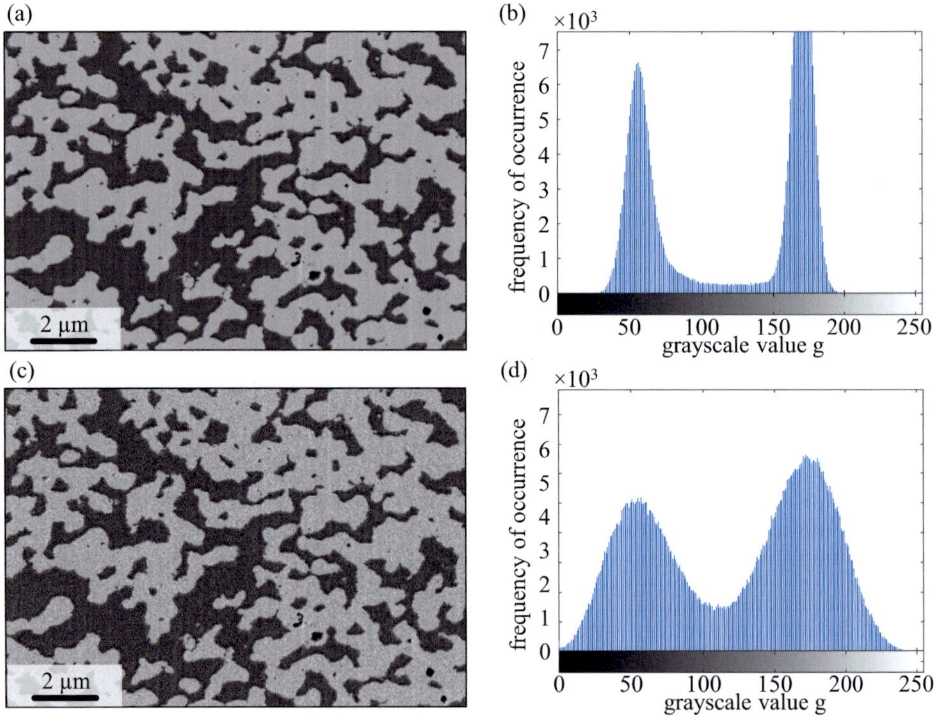

Figure 2.12.: Influence of noise on image quality and histogram: (a) Image of a porous LSCF cathode with a low noise level (epoxy-filled pores in black) with (b) corresponding histogram, where the two peaks are clearly separated. With increasing noise level (c), the peaks tend to overlap (d).

Apart from noise, other artefacts can occur within the FIB/SEM data. Contrast or brightness gradients are often reported in literature. As discussed in Section 2.5.1, they can occur within a single SEM image, e.g. due to shadowing. But brightness and contrast can also change during image acquisition from the first to the last images. This causes a gradient across the slicing direction. The histograms for the individual images will then change. This means that the position of peaks for images recorded deeper within the structure will mostly move to darker greyscale values. Less often they become brighter with increasing depth. A histogram of the entire data set would therefore always be an average over regions with different properties. This also leads to an unclear histogram.

Many methods of correcting this exist. One easy method of removing gradients between the consecutive images is to adjust the peaks in the histogram of each image to the same greyscale values, using simple mathematical operations. There are more advanced and complex methods. One such uses *Discrete Fourier Transform*, which can remove brightness gradients in all directions. A good description of that can be found in [105].

An important tool for processing image data and correcting image defects are filters. In this context, a "filter" is a process that removes some unwanted components or features from an image with the goal of making an image easier to segment later on. From a mathematical point of view, a filter is a convolution. Two common methods of improving the image quality are *median* and *anisotropic diffusion* filters.

The median filter takes a group of pixels and replaces the centre value of this group by the median value of it. This group is defined by a structuring element (also termed *window* or *kernel*), which is usually square or in 3D cubic, but can be any shape. Thus, the median filter removes impulse noise (outlying values, either high or low) and provides localized smoothing. However, if noise is present, a slight blur can result on edges.

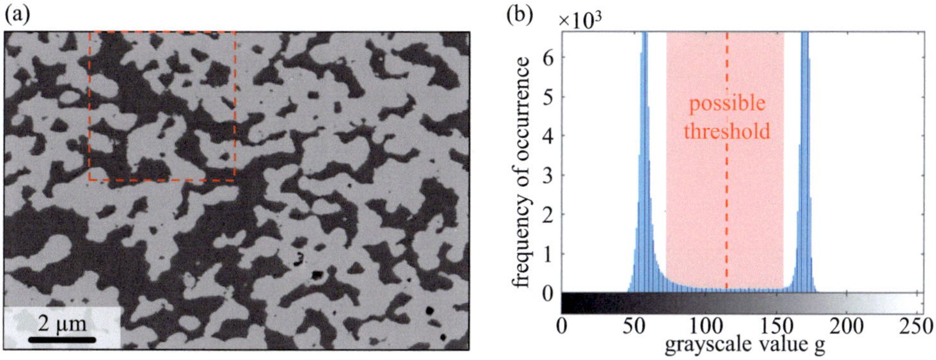

Figure 2.13.: (a) Cross-section of the porous LSCF cathode shown in Fig. 2.12a after applying the anisotropic diffusion filter on it. (b) The peaks in the histogram become narrower and can be separated even better (cf. Fig. 2.12b).

The *Anisotropic Diffusion* filtering method is edge-preserving and able to remove high frequency noise very effectively. It was first proposed by Perona and Malik [106] (therefore also called Perona–Malik diffusion) for multiscale description, enhancement and segmentation of images. The idea is to regard image intensities as particles in 2D or 3D image space. Then the diffusion of particles is equivalent to filtering [105]. A comprehensive description of this method is given in [107]. Figure 2.13 shows an example of this filter, where the image of Fig. 2.12a is shown after filtering. The two phases are clear and smooth, which results in narrower, more defined histogram peaks (cf. Fig. 2.13b).

To obtain a 3D material distribution from the consecutive images, it is necessary to assign each voxel to one of the phases. This process is called *segmentation*. The simplest method is to apply a threshold value. In the case of only two phases, all voxels with a greyscale value above the threshold value are assigned to phase one, while the remaining voxels are assigned to phase two. Figure 2.14a shows the highlighted part of Fig. 2.13a after threshold segmentation with a threshold value of 114. The phases could be adequately separated, since the peaks do not overlap. Lower contrast or higher noise level would lead to segmentation errors, which would require pre- and post-processing of the data. Alternatively, more sophisticated segmentation algorithms can be used, as described in Section 4.6.

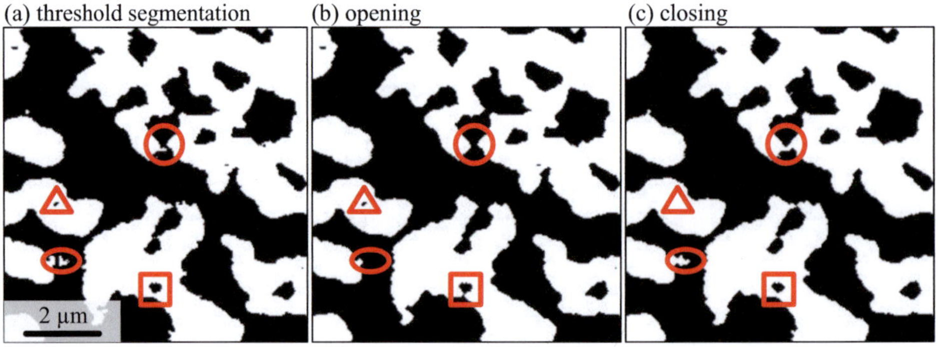

Figure 2.14.: (a) Threshold segmentation of the highlighted part of Fig. 2.13a (threshold value 114). Result of the (b) *opening* and (c) *closing* operation applied to (a). Details of the structure (here surrounded in red) change due to this operations.

Besides greyscale image filters, there are also morphological image processing methods, which are quite common. These are usually applied to binary images, using the relationships between neighbouring pixels. Mathematical morphology is based on the two operations *dilation* and *erosion*. Dilation "grows" objects in binary images (in our case a specific phase is set to "white", while all others are set to "black"), which means that the boundaries increase. The thickening of this phase causes holes within it to appear diminished. Similar to the median filter discussed above, the specific manner and extent of the thickening is controlled by a structuring element [100]. In contrast, erosion "shrinks" or "thins" objects or phases. Thus, holes within this phase seem widened. As with dilation, the manner and extent of shrinking is controlled by a structuring element.

Applying first erosion and then dilation with the same structuring element is known as morphological *opening*. A dilation followed by an erosion is termed *closing*. Figure 2.14 shows an example of the two operations, applied on the threshold segmented image in Fig. 2.14a. Here, the boundaries of the white phase were extended and reduced by 2 pixels.

Figure 2.14 clearly demonstrates that these operations change the appearance of structural details, where some are highlighted in red. The opening process enhances the (black) background pixels. Fine details like small particles and narrow details (smaller than the structuring element) are removed (highlighted as ovals and circles). Thin connections within the structure can be lost (squares). The boundaries of the structure are smoothed with the radius of the structuring element. Closing enhances the (white) foreground pixels, thus thickening structures. Holes smaller than the structuring element are filled (triangles). New connections between closely adjacent regions can appear (ovals and circles). Naturally, for larger structuring elements, these effects increase. Therefore, morphological operations have to be adopted with caution, since they can substantially change structural characteristics.

Sets of pixels can be divided into contiguous groups or parts, by analysing the neighbourhood relationships. Isolated parts can be identified and removed, if required. This constitutes the basis for the *island removal* and *cavity fill* algorithms. Both algorithms can be very useful in processing and analysing the segmented data, as will be explained in Chapter 4.

3. Cells and Samples

This chapter describes cell design, material composition of the analyzed cells, and the different layers. All samples analysed in this thesis - except for the type 3 cathodes - are from FZJ-type cells, which means they are anode-supported cells (ASC, see Section 3.1). The samples from cathodes type 3 are taken from symmetrical cathode cells, as will be described in Section 3.2.

3.1. Anode-Supported Cells (ASCs)

At present, anode-supported cells are the best performing SOFCs. Within this work different types of planar ASCs, as manufactured e.g. by Forschungszentrum Jülich (FZJ), are investigated. The basic design of state of the art ASCs from FZJ is illustrated in Fig. 3.1 and will be described here, together with the manufacturing process of the different layers. This ASC is industrially fabricated on a large-scale and is one of the best performing, most stable cells available, with a very high reproducibility. Reproducibility is extremely important when looking at degradation and microstructural changes during operation, as will be shown in Section 4.9. Many studies reported in literature are based on cells which were manufactured in a lab-scale, "self-made" setup. However, if these cells show a certain scatter in their microstructure due to fabrication, it is hard to be certain if this was a degradation issue or an small-batch production fault. A microstructural flaw caused during fabrication, could be falsely attributed to degradation.

As the name implies, in ASCs the anode provides the mechanical stability for the cell. The anodes in state of the art ASCs are designed to be double-layered with an anode functional layer (AFL) and an anode substrate. Both layers consist of (1) nickel as catalyst and electronic conducting solid, (2) yttria-stabilized zirconia (8YSZ) as ionic conducting solid, and, (3) a pore phase. The main idea behind having two layers is that the AFL provides ideal conditions for the electrochemical reactions inside the anode, while the substrate layer simultaneously provides mechanical stability and enables the transport of reaction species (cf. Section 2.3). The electrochemical reactions inside the anodes mainly take place in the immediate vicinity (some micrometers) of the porous electrode/dense electrolyte interface [109]. Thus a thickness of only a few micrometers is assumed sufficient for the

AFL. The different ends of the layers determine their "ideal" microstructure, which in turn means that both layers should be designed differently.

All single cells characterised here are built on a $50 \times 50\,mm^2$ anode substrates (Ni/8YSZ) with a thickness of 300 to 1500 µm. These porous substrates were manufactured by warm pressing resin-coated powders (*Coat-Mix* process, cf. [110, 111]). They were then pre-sintered at $T = 1285\,°C$. During this sintering the binder burns and leaves pores. The substrate is coated with approximately 7 to 17 µm of AFL and a dense electrolyte layer ($Y_{0.16}Zr_{0.84}O_{2-\delta}$, 8YSZ) of between 7 and 10 µm. The layers are fired together at a temperature of $T = 1400\,°C$. An approximately 7 µm thin $Ce_{0.8}Gd_{0.2}O_{2-\delta}$ interlayer is deposited via screen-printing and sintered at $T = 1300\,°C$ for $t = 3\,h$. This necessary GDC interlayer prevents chemical reactions between LSCF and 8YSZ, which would otherwise form an insulating layer of $SrZrO_3$ [15, 27].

The cathode ($La_{0.58}Sr_{0.4}Co_{0.2}Fe_{0.8}O_{3-\delta}$) is then screen-printed onto the GDC interlayer. After sintering at $T = 1080\,°C$ for $t = 3\,h$, this LSCF cathode layer has a thickness of about 45 µm, while the active area of the working cathode is $10 \times 10\,mm^2$. Two auxiliary electrodes are applied in front of and behind the working cathode (relative to the gas flow direction) to help control the open circuit voltage (OCV). The lateral dimensions of the GDC interlayer are $12 \times 30\,mm^2$ (see Fig. 3.1).

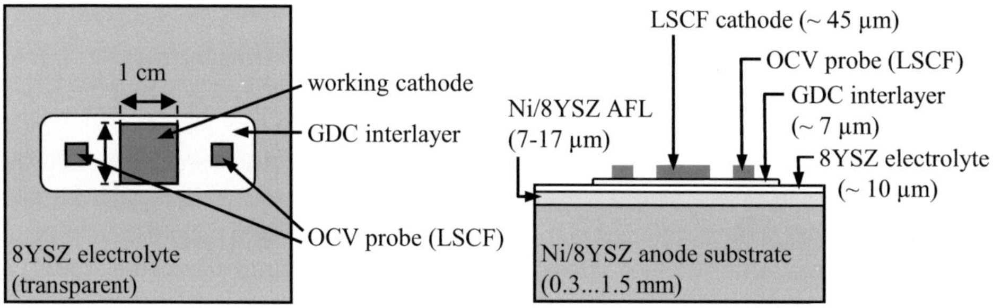

Figure 3.1.: Design of the planar ASCs as applied for the characterisation in this thesis: (Left) top view and (right) side view (not to scale; adapted from [108]).

A cross section of a fractured state of the art ASC from Forschungszentrum Jülich is shown in Fig. 3.2. It covers a part of the porous anode substrate, the AFL, the dense electrolyte, the GDC interlayer, and the porous cathode. Details about its manufacture and the used materials are given in Refs. [111–114].

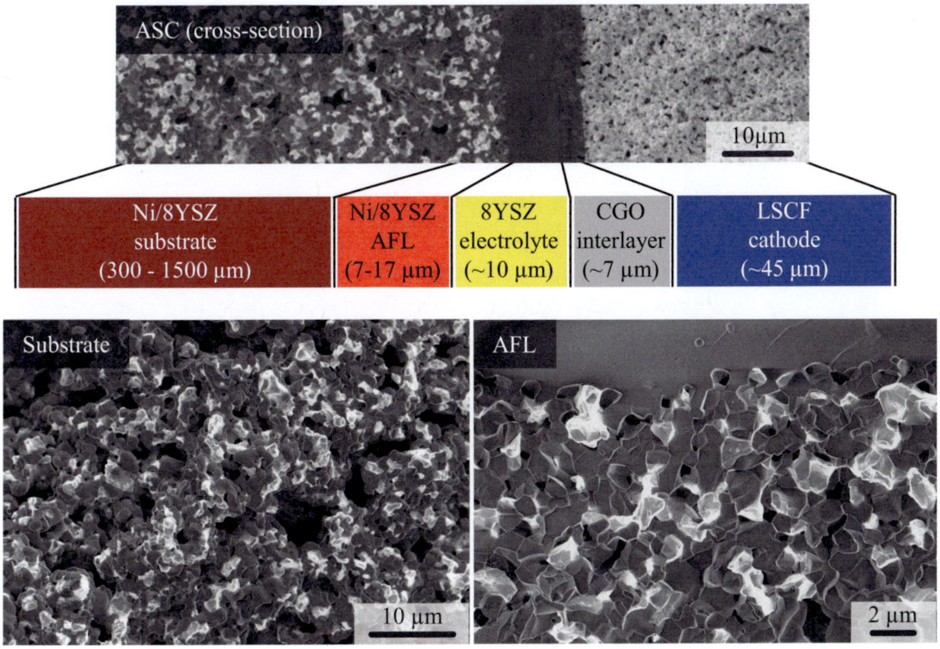

Figure 3.2.: Cross-section of a fractured anode-supported cell showing the porous anode (substrate and AFL) and cathode, the dense electrolyte, and the GDC interlayer. Below the magnified details show parts of the anode substrate (left) and the AFL (right).

All cells were operated at the standard test benches of the Institute for Applied Materials (IAM-WET), see e.g. Refs. [2,8] for more details. A standard weight of 200 g was placed on top of the active cathode area. The reduction of NiO to Ni was performed at $T = 800\,^{\circ}\mathrm{C}$ by an incremental variation of fuel composition from 100 % N_2 to 100 % H_2 and with several hours of settling time. The electrical performance of the ASCs were already evaluated at great detail via electrochemical impedance spectroscopy (see e.g. Refs. [2,37,38,115]).

3.2. Symmetrical Cathode Cells

If a study calls for the specific characteristics of only one of the electrodes, the concept of symmetrical cells can be used. Such cells consist of the same electrode on both sides of the dense electrolyte. This thesis investigates symmetrical cathode cells consisting of a LSCF cathode with a GDC interlayer on both sides of an 8YSZ electrolyte (see Fig. 3.3). Since the cathodes should have approximately the same thickness as in "technically relevant" ASCs, the mechanical stability must be provided by the dense electrolyte which is therefore thickened. This symmetrical design has the advantage that the influence of the cathode can be investigated in isolation from the anode. In ASCs the anode and cathode processes are sometimes hard to separate (e.g. the anode processes overlap the cathode processes in the high frequency range of the impedance spectra). This is especially true if looking at the electrochemical processes of the cells.

The symmetrical cathode cell design is illustrated in Fig. 3.3, with an additional cross-section of one cathode. The manufacturing process is very similar to that of the ASCs (see Section 3.1). First a $25 \times 25\,\mathrm{mm}^2$ 8YSZ electrolyte with a thickness of $180\,\mu\mathrm{m}$ (from Itochu, Japan) is purified. Then an approximately $7\,\mu\mathrm{m}$ thin GDC interlayer is deposited in the center of the substrate via screen-printing on both sides. These interlayers ($11 \times 11\,\mathrm{mm}^2$) are sintered at $T = 1300\,°\mathrm{C}$ for $t = 3\,\mathrm{h}$. LSCF cathode layers with a thickness of approximately $60\,\mu\mathrm{m}$ were deposited on top of the GDC layers. As with the ASCs, the active area of the working cathodes are $10 \times 10\,\mathrm{mm}^2$. The last step of the manufacturing process is sintering the LSCF cathodes. This was done at different temperatures for different cells, as will be explained in the next section for these type 3 cathodes. The electrochemical characterisation of the symmetrical cells here analysed was done in a study project which was supervised in the context of this work [116].

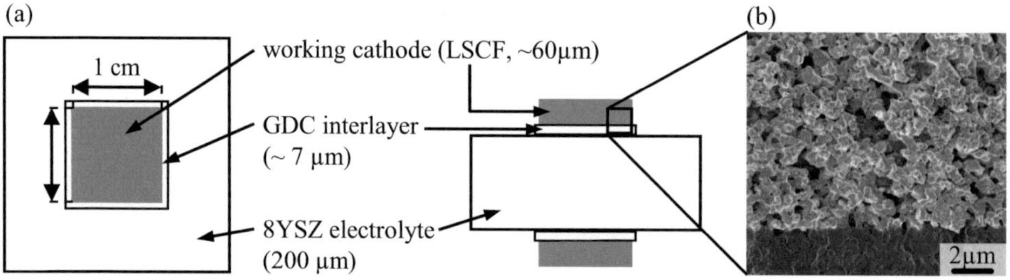

Figure 3.3.: Design of the symmetrical cathode cells applied for the characterisation in this thesis: (a) top view and side view (not to scale). (b) Scanning electron micrograph of a fractured cell cross-section showing one of the porous cathodes and the GDC interlayer.

3.3. Cathode Samples and Aim of Studies

In this section details of the different types of cells are provided and the aims of the investigations are introduced. Three different types of cathodes are analysed, which are referred to as type 1, 2 and 3. All three sample types are LSCF cathodes from FZJ with the same nominal composition ($La_{0.58}Sr_{0.4}Co_{0.2}Fe_{0.8}O_{3-\delta}$). The main differences between the samples and the focus of the characterisation of these cathodes are explained separately in the following, and summarized in Tab. 3.1.

Table 3.1.: Cell design, information on sintering and history of all cathode samples characterised within this work.

Sample	Cell Design	T_{sint}	Sintered at	Treatment before FIB Analysis
Type 1_{AP}	ASC	1040 °C	FZJ	"As prepared"
Type 2_{AP}	ASC	1080 °C	FZJ	"As prepared"
Type 2_{600}	ASC	1080 °C	FZJ	> 1000 h at 600 °C
Type 2_{750}	ASC	1080 °C	FZJ	> 1000 h at 750 °C
Type 2_{900}	ASC	1080 °C	FZJ	> 1000 h at 900 °C
Type 3_{960s}	Sym.-Cat.	960 °C	KIT	"As prepared"
Type 3_{1030s}	Sym.-Cat.	1030 °C	KIT	"As prepared"
Type 3_{1080s}	Sym.-Cat.	1080 °C	KIT	"As prepared"
Type 3_{1200s}	Sym.-Cat.	1200 °C	KIT	"As prepared"

Type 1

Type 1 cathode is an LSCF cathode manufactured at FZJ, where the main difference to the "standard manufacturing process" described in Section 3.1 is that the cathode was sintered at 1040 °C (the standard sintering temperature for LSCF cathodes is 1080 °C). Another difference is that the GDC interlayer was not deposited via screen-printing, but with physical vapour deposition (PVD, see e.g. Ref. [29] for more details). Therefore the GDC interlayer is dense and much thinner (less than 0.5 µm). This cathode was used for the first reconstruction conducted during this work. The reconstructed volume is divided into three sub-volumes and the differences between these volumes will be investigated in Section 4.9, alongside the question of whether these volumes are representative. Moreover, the type 1 cathode is used to investigate accuracy aspects (Section 4.8). Details of their microstructural characterisation and simulation have been presented in various publications [117–119].

Type 2

The type 2 cathodes are from four identically manufactured ASC cells from FZJ. Ergo, their microstructures after manufacturing ("as prepared") are nearly identical. This cathode type represents the state of the art LSCF cathode from FZJ, so the manufacturing process was exactly as described in Section 3.1. The aim of this study is the observation of microstructural changes during operation. Therefore three cells were operated for more than 1000 h at temperatures of $T = 600$ °C, 750 °C and 900 °C. The fourth cell was not operated before reconstruction and kept as reference ("as prepared").

The long-term electrochemical characterisation of the cells was conducted and reported by Cornelia Endler-Schuck in her PhD thesis [27]. The ohmic and polarization losses of electrolyte, anode and cathode were separately identified by high-resolution impedance studies on these ASCs over the entire operating time. The cells showed a degradation, which means that the performance became worse over time. Thereby the degree of degradation depends on the operating temperature, see Ref. [39]. However, it was not known if this degradation is mainly due to microstructural change, or due to a diminishing of material characteristics over time. Therefore, the LSCF cathode microstructures are quantified with FIB tomography before and after the impedance studies here (cf. Section 4.9). This enables, for the first time, the use of actual values for porosity, surface area and tortuosity in the calculation of specific material parameters for describing the material characteristics of LSCF cathodes (see Ref. [39]). Thus it is possible to quantify both the material and microstructural contributions to cathode degradation.

Since this cathode type represents the state of the art cathode, most of the studies on accuracy aspects were performed with the "as prepared" sample (type 2_{AP}).

Type 3

As already mentioned, the type 3 cathode samples were not taken from ASCs, but from symmetrical cathode cells. The manufacture of these cells was as described in Section 3.2, except for the sintering temperature of the LSCF cathodes. The microstructure of the porous electrodes is strongly related to the manufacturing process. High-temperature sintering is required for this, which has a big influence on the resulting microstructure. The most crucial parameters are the sintering time (t_{sint}) and temperature (T_{sint}). With increasing time or temperature, the particles (and pores) of the structure will grow, i.e. the microstructure will get coarser. The aim of this study is a detailed investigation of the sintering temperature on the LSCF microstructure. Thus, different cells were sintered at different temperatures. To that point, all other manufacturing steps were identical. The four different cells investigated here were sintered for 3 h at $T_{\text{sint}} = [960, 1030, 1080, 1200]$°C. The electrochemical characterisation of these samples were completed in a study project which was supervised in the context of this work [116].

4. 3D Reconstruction of SOFC Electrodes

This chapter is the main part of this work. It deals with the cathode reconstructions by FIB tomography and the following quantifications and describes the methods and procedures needed for an exhaustive reconstruction and quantification of SOFC electrodes. A large amount of work was conducted to minimize the errors and to ensure the reconstruction data were correct (see Sections 4.7 and 4.8).

In the first section an overview is given of the worldwide developments and the current state of SOFC electrode reconstruction via FIB tomography. Afterwards the sample preparation is described in Section 4.2. Next, the multistep procedure of FIB tomography and data acquisition is analysed in Section 4.3. This is followed by a brief description of image3D Reconstruction of SOFC Electrodes processing methods in Section 4.4. An advanced reconstruction method of using two detectors for the data acquisition is discussed in Section 4.5. Algorithms for the adequate segmentation of three-dimensional image data are presented in Section 4.6. This is followed by a thorough synopsis of the microstructural parameter estimation methods (Section 4.7). The chapter ends with an analysis of FIB tomography's accuracy aspects and links it to the results obtained from the investigated cathodes. Most of the aspects and methods described here have been previously published in peer-reviewed journals and in conference proceedings [101, 104, 117, 118, 120, 121]. Most of these studies were recently enhanced: progress had led to larger reconstructions with fewer artefacts.

4.1. State of the Art: SOFC Electrode Reconstruction via FIB Tomography

It has been recognized for decades that the importance of microstructure is a key for understanding the relationship between processing, microstructure and performance. However, a detailed 3D quantitative structural analysis was hindered for many years [122]. This situation has changed rapidly, with many improvements made in the field of 3D reconstruction techniques in the last years. FIB tomography (and to a lesser extent X-ray tomography, cf. Section 2.5) has been widely used for SOFC electrode reconstruction. This section provides an overview on the worldwide activities of SOFC electrode reconstruction via FIB tomography. It discusses both cathode and anode reconstructions, even though only cathode reconstruction results are presented here. This is because methods for both electrodes were developed within the dissertation project. Advancements prior to this work are described, as well as the progress achieved by contemporary groups during the span of this work. For the sake of clarity, a review of only selected publications which have helped shape the field are given.

FIB tomography is still a relatively fresh technique. The first reconstruction of a SOFC electrode was reported in 2006 [86]. When this PhD project commenced in 2009, only three other articles about FIB tomography on SOFC electrodes had been published. Since then, the interest in this technique for SOFC research has increased rapidly. Consequently, numerous research groups worldwide have started to use FIB tomography for their SOFC investigations (e.g.: Barnett/Northwestern University, US [86, 123–125]; Shikazono/Tokyo University, Japan [126–129]; Eguchi, Iwai/Kyoto University, Japan [97,130–133]; Wachsman/University of Florida, US [134, 135]; Brandon, Shearing, Atkinson/Imperial College, UK [102, 136, 137]; Jørgensen/ DTU, Denmark [138, 139]; Holzer/ZHAW Winterthur, Swiss [140–142]); Laurencin, Vivet/CEA, France [143, 144] and many others).

Recently, a number of good reviews on this topic were published, covering the progress of the technique itself (e.g. [91, 122, 145]) and SOFC related reconstruction studies in the last few years ([83, 84]). However, each of these groups developed their own methods, making it hard to compare results from the different groups, as will be discussed later.

The first 3D reconstruction of an SOFC electrode was reported by Wilson *et al.* [86] in 2006. Based on 82 consecutive images of a Ni/YSZ anode with a distance between the images of about 50 nm (in a later publication [103] it was reported as 44 nm), the microstrcuture was characterised. The segmentation process was done manually, which is not only extremely labour intensive, but also unreliable with low-reproducibility, as manual segmentation is ultimately subjective [146]. On the other hand, an automated segmentation was supposedly not possible, hampered by weak contrast between the phases. It wasn't then common to fill the pores of the structure with contrast-enhancing epoxy resin. It was therefore hard to automatically separate pores from material (cf. Section 2.5.2). However, from the resulting reconstruction different properties were obtained including surface area density, TPB density, connectivity and also the tortuosity of the pore phase (by performing simulations based on the reconstructed structure with a commercially available finite element modelling tool). Thereby the tortuosity was calculated along all 3 axes, obtaining values between 1.9 and 2.2. This first reconstruction already demonstrated the enormous potential of this technique, even though a volume of approximately $5 \times 3.4 \times 6\,\mu m$ was reconstructed, which is arguably small for providing sufficient statistics. Based on the same experimental data Wilson *et al.* [103] later published methods for analysing the interconnectivity of the phases and the TPB calculation.

In 2007 Gostovic *et al.* [134] reported the first reconstruction of a pure LSCF. From three symmetrical cells which had been sintered at different temperatures (850, 950 and 1100 °C) three pillars from each sample were reconstructed with a total volume of 399, 227 and 439 μm^3, respectively. The resolution of the SEM images gave a 3 nm pixel size, which was much smaller than the slicing distance of 20 nm. Parameters like porosity fraction, surface area density, average pore diameter and (geometrical) tortuosity of the pores were calculated, but it is unclear whether the nine analysed volumes were large enough to ensure reliable results. The tortuosity calculations were based on the ratio between the distance travelled along the centroid of a pore to the electrode thickness. This method provides a useful metric for comparing the 3D pore distributions between different samples, but does not account for pore constrictions influencing the effective diffusivity [84]. However, average tortuosity values between 2.5 and 2.8 were obtained. The extraordinarily high standard deviations for the calculated parameters indicate that their reconstructed volumes were not large enough. The calculated porosity is especially questionable. It reports increasing porosity with increasing sintering temperature, which contradicts the physics of sintering. Also the stated explanation is not satisfying, it claims that this is the result of a lower relative resolution for the structure sintered at 850 °C compared to the cell sintered at 1100 °C. Another explanation might be that the structure was imperfectly segmented, since the pores were not filled with epoxy resin. However, this is purely speculative and can only be observed when inspecting the segmented data (which are not given in the paper). Nevertheless, they were the first to investigate the effects of processing on electrode microstructure.

Two years later, Smith *et al.* [135] from the same group reported the reconstruction of a single-phase LSM cathode on a YSZ substrate. Although the volume size of the reconstruction was not reported, they mention that about 30 images were taken per sample with a slice distance of 50 nm. Thus the reconstruction volume would only have been around 1.5 µm thick, which is, again, arguably small for providing sufficient statistics. The goal of the study was to investigate the effect of sintering on the microstructure and the electrochemical performance. As in their previous publication, the TPB contact length at the cathode/electrolyte interface was reported per unit area for this single-phase cathode.

In 2009 Wilson *et al.* [123] had also been the first to publish a reconstruction of a LSM/YSZ composite cathode. From 242 images (slice distance of 53.3 nm) a total volume of 685 μm^3 could be achieved. Again the TPB density and the connectivity were calculated, giving the result that only 33 % of the total TPB belonged to a continuous network, which is considered to be a precondition for electrochemical activity. It should be noted that the pores were now filled with epoxy resin (in contrast to their previous work).

Shearing *et al.* [102] calculated volume fractions and the TPB length from a volume they extracted using the FIB lift-out technique prior to sectioning. They thereby improved contrast between the phases and reduced shadowing. However, the FIB lift-out technique is far too labour intensive to become a standard method. Moreover, the volumes to be "liftedout" are relatively small (here, only 1.2 µm thick), which is too small to be representative. However, they showed that the image quality clearly benefits when the investigated volume is exposed.

Kishimoto *et al.* [133, 147] calculated the tortuosity based on random walk methods. Similar to the method of Gostovic [134] this method does not account for constrictions within the transport paths. However, the presented values for the pore phase inside a Ni/YSZ anode are in broad agreement with the data published in [86] and [134] (1.77 - 2.16 in [133] and 1.70 - 1.95 in [147]). The tortuosities of the solid phases are much higher, with values of between 6.09 and 17.1 (5.90 - 8.85 in [147]) for the YSZ, and 5.66 to 14.7 (6.47 - 6.91 in [147], respectively) for the Ni phase. In [147] they calculated the tortuosity in transport direction and in the other two directions, resulting in extremely diverse values with a difference of up to one decade (for Sample 1 the tortuosity of nickel is 64.4 in y-direction and 6.47 in z-direction, while there is no connected path in x-direction). The author concluded that the sample volumes of 1013 to 1664 μm^3 had been too small and that further studies would require larger volumes. This might be true, but the different physical dimensions of the reconstructed volume (e.g. volume of sample 1: $25.4 \times 10.4 \times 4.56\,nm^3$) and of the voxels (26.6 nm in x- and y-direction, but 60 nm in slicing direction z) might also play a role.

The same approach to tortuosity calculation presented in [133] is used in Iwai *et al.* [97]. They come from the same research group and compared their results with tortuosity values calculated by a Lattice Boltzmann Method (LBM) of transport simulation. Both techniques were found to be in good agreement for the Ni/YSZ anodes investigated here. Unfortunately the computational expense of the two methods were not reported. They also compared two methods of TPB density calculation. One is based on a *centroid* method and the other on an alternative volume expansion method. Again, both methods were found to give good agreements. As in their previous publication, they reconstructed three Ni/YSZ anodes, and it seems like they might have been the same data sets.

The same image data are used by Shikazono *et al.* [126]. Again the importance of accurately determining the TPB length is addressed. It was shown that the error depends on the image data resolution, among other things. Using the example of two overlapping spherical particles (one is Ni and the other YSZ), where the TPB would be a circle at the connection between the two particles, they show that the TPB would be overestimated by up to 80 % depending on the method, the resolution and the overlap of the particles. Three methods for calculating the TPB are presented: (1) the simple summation of the identified TPB edges, (2) the *midpoint method* which defines the connection length of the midpoints of the TPB edge segments and (3) the *centroid method* which gives the total distance between the centroids of the triangles, defined by the neighbouring midpoints of the edge segments. The latter especially has been shown capable of accurately approximating the TPB length.

In 2010 Jørgensen *et al.* [138] presented a very sophisticated method for accurately segmentating 3D image data. They were the first to have really focused on the segmentation process.

Although segmentation is a key point for obtaining reliable results from tomography data, not much is reported in literature concerning SOFC electrode reconstruction. The segmentation procedure was either done by hand [86, 123], semi-automatically [97, 125, 126] or reported as "grey level-based thresholding" [144] or based on "brightness of image" [147].

In the same year, Jørgensen *et al.* [148] also presented sophisticated methods for calculating interfacial areas and TPB curve length. Some years later they published an interesting paper about the accuracy of TPB length calculations from tomography data [139], which presents suitable methods for the calculation of the TPB.

Vivet *et al.* [144] also calculated the TPB length (by simple edge summation), interfacial areas and tortuosity of a Ni/YSZ anode. The tortuosity was calculated by solving the Laplace equation within the investigated phase. As the resolution of the 115 SEM images was much higher compared to the slicing resolution, the data set consists of very anisotropic voxels ($10.4 \times 13.2 \times 100\,\mathrm{nm}^3$). Before calculating the tortuosity and TPB length, the data set was re-sampled to obtain approximately a billion, $10 \times 10 \times 10\,\mathrm{nm}^3$ cubic voxels. However, such a large up-sampling could give errors, which was not discussed. This might also be one reason why the tortuosity of nickel in slicing direction (z-direction) was twice as large ($\tau_{\mathrm{Ni}} = 6.24$) as in the other two directions ($\tau_{\mathrm{Ni}} = 3.04$ and 3.18, respectively).

Nelson *et al.* [149] compared the microstructural quantification of X-ray nanotomography to FIB tomography by reconstructing a LSM/YSZ composite cathode. The pixel size of the SEM images was 10 nm, but in order to compare the results the FIB data were down-sampled to a pixel size of 30 nm, to match with the data obtained by X-ray. The results of both techniques were in good agreement (which is particularly true for the volume fractions and the phase connectivity), but the TPB density calculated from the X-ray tomography data were about 20 % smaller.

In a similar study on a Ni/YSZ anode support layer, Quey *et al.* [150] found that X-ray nanotomography underestimates pores smaller than about 1 µm while overestimating the size of pores larger than about 1.5 µm. In previous work using X-ray nanotomography [85] the same group had proposed that when investigating the anode substrate layer, a volume in the $(40 - 50\,\mu\mathrm{m})^3$ range would be representative. It is worth noticing that they observed the same exact position with both techniques.

In the last few years, FIB tomography has often been used to study whether and how the electrode microstructure changes during operation. Therefore, the method itself is often not the main focus of the publication and the reader is referred to previous publications. For example, Cronin *et al.* [125] studied high temperature annealing of Ni/YSZ anodes by comparing a non-annealed sample against a sample annealed for 100 h in a 4 % $H_2 - H_2O$ 3 % - 93 % Ar mixture at 1100 ° °C. They reported a 50 % reduction of TPB for the annealed sample, which they attributed to increased pores. They did not observe nickel coarsening, which they think was countered by the YSZ skeleton. However, many other studies have observed nickel coarsening at operating temperatures (e.g. [151]).

Table 4.1.: Main parameters of 3D SOFC electrode reconstructions performed via FIB tomography as published in literature (the list is far from exhaustive). *Note that each of the volumes from Gostovic *et al.* [134] are divided in three sub-volumes.

Author, Year	Citation / Data set	Type	Volume / μm^3	Pixel Size / nm	Image Dist. / nm	No. Images / -	No. Voxel $\times 10^6$
Wilson 2006 (2009)	[86] ([103])	Ni/YSZ	~109	13.9	50 (44)	82	~12.8
Gostovic 2007	[134]/850 °C	LSCF	399 (/3*)	3	20	N/A	N/A
	[134]/950 °C	LSCF	277 (/3*)	3	20	N/A	N/A
	[134]/1100 °C	LSCF	439 (/3*)	3	20	N/A	N/A
Smith 2009	[135]	LSM	N/A	N/A	~50	~30	N/A
Wilson 2009	[123]	LSM/YSZ	685	N/A	53.3	242	N/A
Shearing 2009	[102]/1	Ni/YSZ	75	19×26	30	40	~6.9
	[102]/2	Ni/YSZ	722	20	19	175	~95
Iwai 2010	[97]/1-3	Ni/YSZ	1349-1965	26.5-26.6	60-72	79-106	~32-45
Shikazono 2010	[126]	Ni/YSZ	972	26	62	N/A	~23.2
Shearing 2010	[152]	Ni/YSZ	50.5	20	15	100	~8.4
Matsui 2010	[130]/1,2	Ni/YSZ	3150-3368	27.9	61	191-213	~66-71
Vivet 2011	[144]	Ni/YSZ	967	10.4×13.2	100	115	~70.5
Matsuzaki 2011	[128]/1	LSCF	3392	13.96	59.18	97	294
Cronin 2011	[125]/1,2	Ni/YSZ	914-1121	25	50	N/A	~29-36
Kanno 2011	[127]/A	Ni/YSZ	2424	37.2	74.5	115	~23.5
	[127]/B	Ni/YSZ	3905	37.2	61.7	175	~45.7
	[127]/C	Ni/YSZ	17399	55.8	74.7	194	~74.8
Matsui 2012	[131]/IS,A,B	Ni/SCZ	1310-1402	44.67	97-105	96-100	~6.5-6.8
Delette 2013	[143]	LSCF	1120	10	10	800	1120
Yakal-Kremsky 2014	[153]	NNO	1075	20	40	N/A	~67.2
Chen 2015	[137]	LSCF	N/A	12.5	12.5	N/A	N/A

Holzer *et al.* [141] have also observed nickel coarsening in Ni/CGO anodes. They observed nickel growth in both dry and humid atmosphere using FIB tomography with a resolution of 15 - 25 nm in the SEM images and 12 nm image distance. They report a very high growth rate for nickel in the first 200 h which decreased significantly after 1000 h in a humid environment and showed a low but constant growth rate under dry conditions. In addition to these findings an advanced method for the segmentation of the tomography data based on a region growing process was presented, similar to Ref. [138].

Later, Matsui *et al.* [131] studied the microstrcutural change of a Ni–10Sc1CeSZ anode from a cathode-supported tubular cell. At a resolution of 44.67 nm pixel size and 97.2 to 104.9 nm image distance, they observed three volumes (1310 - 1402 μm^3), one being the reference (as prepared) and the other two from cells discharged for 2500 and 6500 h, respectively. They observed a strong aggregation of nickel particles in the anode, as well as a migration. After 6500 h, the nickel particles were significantly diminished in the vicinity of the anode/electrolyte interface, having most likely migrated towards the current collector. This effect resulted in a significant decrease in TPB density.

In 2014 Kishimoto *et al.* reported the reconstruction of an infiltrated anode consisting of a GDC scaffold through which Ni nano-particles were infiltrated. Three reconstructions were made, showing the GDC scaffold without infiltration and two structures differently infiltrated (one-time infiltration and ten-time infiltration). The size of the nickel particles was reported as between about 100 and 350 nm for the two infiltrated structures. Unfortunately the resolution of the data set was not reported, making it difficult to evaluate the result. Nevertheless the TPB density is reported to be ten times larger in these infiltrated structures compared to conventional anodes. However, the long-term stability of such structures under operating temperatures is highly problematic, but was not investigated here.

Besides the discussed studies, many other reconstructions via FIB tomography were reported during the last years, for all kinds of material systems. Available information on published FIB tomography data sets are listed in Table 4.1 (the list is far from exhaustive). The table allows a comparison of analysed volume, pixel size, image distance, total number of images and the number of voxels. Between 30 and 242 images were used to reconstruct volumes of between 50.5 μm^3 and 17 399 μm^3. Thereby the resolutions of the SEM images range from 3 nm pixel sizes up to 55.8 nm, while the distance between the consecutive images ranges from 10 nm up to 105 nm. The reconstructions with larger volumes often show a relatively low resolution (e.g. [127]: volume of 17 399 nm^3 with a resolution of $55.8 \times 55.8 \times 74.7\ nm^3$), while the reconstructions performed at very high resolutions normally only assay a relatively small volume (e.g. [134]: resolution of $3 \times 3 \times 20\ nm^3$ and a volume of 299 nm^3 for sample 2). This is also shown in Fig. 4.1, where the volumes (blue) and the resolutions (number of voxels per μm^3) for some published reconstructions are shown. Reliable results require appropriate resolution and a large enough volume. Hence the trade-off between resolution and volume is of particular importance. Sometimes, however, the resolution was drastically down-sampled for 3D simulations. A high resolution and a large volume will lead to a large amount of data, which can be difficult to handle. The data sets in Table 4.1 mostly consist of about 6.5 to 300 × 10^6 voxel (except for Ref. [143]).

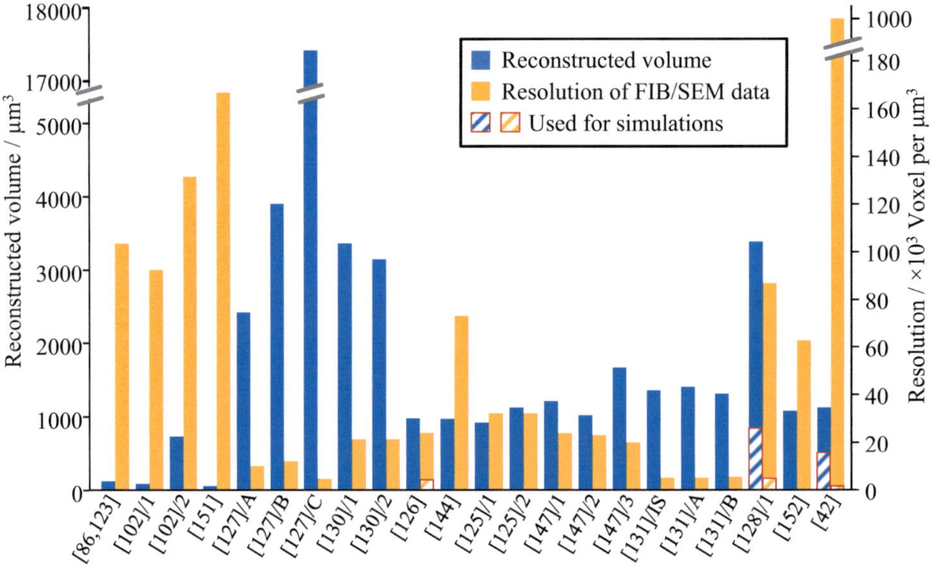

Figure 4.1.: Reconstructed volumes (blue) and resolution (number of voxels per µm^3) of reconstructions published in literature (the list is far from exhaustive).

Despite the large number of publications, many aspects of FIB tomography still remain open, especially with respect to data accuracy and reliability of the methods and results. For example, most publications do not discuss whether their reconstructed volume is large enough to provide appropriate statistics. The same holds true for the necessary resolution of the image data. Moreover, as can be seen in Table 4.1, the pixel size and the image distance often significantly deviate from each other (e.g. Refs. [134, 144]). Nevertheless, neither the effect of such an anisotropic resolution nor the influence of re-sampling if applied to obtain a cubic voxel dataset is discussed. It is also rarely reported that possible sources of error can be introduced during the multi-step FIB reconstruction procedure.

Most of the reconstructions aimed to quantify the porous electrode microstructure and calculate some characteristic microstructural parameters. But since the methods, algorithms and definitions used to calculate these parameters are not consistent across literature, it is difficult to compare results from different groups. Furthermore, most groups fabricate their own cells and these cells are operated differently prior to the reconstruction. Nevertheless, some of the values published in literature are summarized in Table 4.2 for Ni/YSZ anodes and in Table 4.3 for LSCF cathodes.

Table 4.2.: List of microstructural parameters obtained by three-dimensional Ni/YSZ anode structure quantification, as published in literature.

Author	Wilson 06	Wilson 11	Cronin	Iwai	Vivet	Shearing	Delette [143]	
Citation	[86]	[154]	[125]	[97] / S2	[144]	[155]	Substrate	AFL
Group/Place	Northwestern, US			Kyoto, Jap.	CEA, Fr.	Imp. Coll., UK	CEA, Fr.	
Technique	FIB	FIB	FIB	FIB	FIB	X-ray	X-ray	X-ray
Volume / μm^3	109	498	914	1965	967	681	67392	12960
Voxel per μm^3 / $\times 10^3$	11.5	32.5	32	22.8	72.8	93.9	64	64
Vol. % Ni	25.9	33	27.8	25.3	26	61.6	21.5	28
Vol. % YSZ	54.6	46	54.1	25.1	33	19.7	31.5	49
Vol. % Pore	19.5	21	18.1	49.6	41	18.6	47	23
a_{Ni} / μm^{-1}	3.00	N/A	1.57	0.90	2.33	N/A	N/A	N/A
a_{YSZ} / μm^{-1}	1.60	N/A	2.61	1.89	4.24	N/A	N/A	N/A
a_{Pore} / μm^{-1}	2.40	N/A	1.77	2.04	4.27	N/A	N/A	N/A
$a_{Ni/Pore}$ / μm^{-1}	N/A	N/A	0.36	N/A	1.18	N/A	N/A	N/A
$a_{Ni/YSZ}$ / μm^{-1}	N/A	N/A	1.21	N/A	1.15	N/A	N/A	N/A
$a_{YSZ/Pore}$ / μm^{-1}	N/A	N/A	1.40	N/A	3.09	N/A	N/A	N/A
d_{Ni} / μm	N/A	N/A	N/A	N/A	N/A	N/A	1.2	0.83
d_{YSZ} / μm	N/A	N/A	N/A	N/A	N/A	N/A	1.1	0.73
d_{pore} / μm	N/A	N/A	N/A	N/A	N/A	N/A	2.6	0.94
l_{TPB} total / μm^{-2}	4.28	3.97	3.37	2.56	7.4	17.9	N/A	N/A
l_{TPB} connected / μm^{-2}	N/A	3.85	2.6	20	N/A	10.3	N/A	N/A
l_{TPB} isolated / μm^{-2}	N/A	N/A	0.43	20	N/A	N/A	N/A	N/A
τ_{Ni}	N/A	N/A	N/A	N/A	9.2-38.9	N/A	N/A	N/A
τ_{YSZ}	N/A	1.8	N/A	17.4	3.2-4.4	N/A	N/A	N/A
τ_{Pore}	2.1	4.1	5.51	2	N/A	N/A	N/A	N/A

Table 4.3.: List of microstructural parameters obtained by 3D LSCF cathode structure quantification published in literature. Note, that the different groups used different methods to calculate the parameters.

Author	Gostovic [134]		Matsuzaki	Delette	Chen [137]	
Data set	850 °C	1100 °C	[128]	[143]	900 °C	1200 °C
Group/Place	Florida, US		Tokyo, Jap.	CEA, Fr.	Imp. Coll., UK	
Technique	FIB	FIB	FIB	FIB	FIB	FIB
Volume / μm^3	399	439	3392	1120	N/A	N/A
Voxel Size / nm^3	$3 \times 3 \times 20$		N/A	10^3	12.5^3	
Vol.% LSCF	78.3	67.6	N/A	61.7	53.1	84.8
Vol.% Pore	21.7	32.4	N/A	39.3	46.9	15.2
a_{LSCF} / μm^{-1}	9.3	7.2	6.765	N/A	N/A	N/A
d_{LSCF} / nm	N/A	N/A	N/A	910	470	1700
d_{pore} / nm	140.9	268	N/A	620	N/A	N/A
τ_{LSCF}	N/A	N/A	N/A	1.75	~1.4	~1.1
τ_{Pore}	2.5	2.8	N/A	N/A	~1.4	~2.4

4.2. Sample Preparation

A main advantage of FIB tomography is the good contrast between different materials provided by SEM, as described in Section 2.5. To use it to full advantage and to ensure accurate results, careful sample preparation is extremely important. Moreover, an optimal preparation can save significant time and efforts later on during image processing and segmentation.

A segment (approx. 5×3 mm^2) is vacuum infiltrated with a low-viscosity epoxy to fill the open pores. As already mentioned in Section 2.5, this has various advantages: it strengthens the porous microstructure during FIB/SEM treatment and allows for better planar sectioning. It would otherwise be difficult to decide which areas of the image are in the cross-sectional plane, and which are behind this plane and behind pores. Additionally, it supports the separation of the phases, since the resin only emits very low SE and BSE signals. The epoxy-filled pores appear black in the SEM images, which is easy to detect. Furthermore, when more than one solid phase is present, it allows to focus on the contrast between these solid phases. For example, when analysing Ni/YSZ cermet anodes it can be challenging to find a good contrast between Ni and YSZ. Infiltration is an essential precondition to optimize image quality and allow for a (semi-)automated segmentation of the consecutive images, which will be discussed in Section 4.6.

(a) (b)

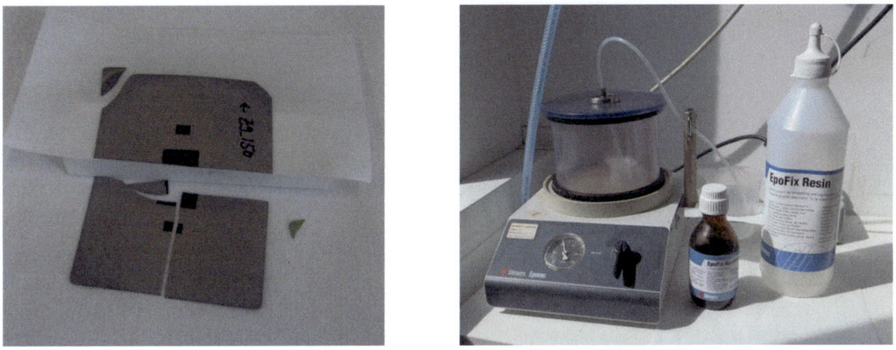

Figure 4.2.: (a) The cell is broken into small pieces. (b) One piece is placed in a vacuum chamber and the epoxy is sucked through a hose and fills a small cup containing the sample.

When a porous structure has not been infiltrated, material which is sputtered away from the sample can re-deposit in the pores. Another negative effect mitigated by epoxy infiltration is the curtaining-effect (or streaking) [102], caused by the spreading of the ion beam. This results in an unsteady removal of material and leads to vertical streaking on the SEM images. The main problem associated with infiltration is air bubbles inside the epoxy [120]. The edges of these bubbles can be charged during FIB/SEM treatment and appear in a similar greyscale as the material phase. The presence of air bubbles can become really challenging for image processing and software-assisted segmentation. In order to minimize this effect, the process of infiltration was optimized. Several different epoxies were tested; Epofix from the Struers company gave the best results. It consists of a resin and a hardener, which are mixed in a ratio of 25 : 3 (by weight). The sample is placed inside a small vacuum chamber in an vacuum impregnation unite from Struers (cf. Fig. 4.2b) and the epoxy is sucked through a hose and fills a small cup containing the sample. The chamber is so configured that the pressure doesn't exceeded roughly 100 mbar. This pressure limitation is set because the high vapour pressure of the solvent would lead to frothing of the resin at lower pressures [98]. It was found that the following procedure significantly lowers the quantity of bubbles: before filling the sample with epoxy, the vacuum chamber should be evacuated with about 100 mbar. Then it is held at this pressure for some minutes after infiltration, during which the bubbles can leave the epoxy. Afterward, the pressure should be slowly increased to ambient pressure, whereby the epoxy is depressed inside the small pores. With this procedure it was found, that in almost all cases bubbles were no longer a problem. With short holding times a drastic increase of bubbles was observed.

The cured sample is cut to approximately $4 \times 10\,\text{mm}^2$, removing excess epoxy. This was previously done using a diamond saw, but grinding with a coarse polishing paper was found to be much faster and less expensive. The sample is finished by polishing with SiC polishing papers down to a finish of < 1 µm. Grinding and polishing are done with two cross-sections perpendicular to each other, which gives a clear edge. This enables FIB/SEM sectioning right from the edge of the sample, as explained in the next section. If there is a problem with delamination of the electrode from the electrolyte, a more stable binding can be achieved by gluing together a pair of infiltrated and cured electrodes face-to-face to produce a "twinned" sample (cf. Refs. [104, 118]). Delamination is especially likely in cells operated for a long time. Twinned samples have the additional advantage that two (or more) regions can be polished and prepared at the same time, but this requires more effort.

After polishing, the sample is mounted on a SEM stub using a conductive Ag lacquer. A thin gold coating layer ($\sim 20 - 300\,\text{nm}$) is sputtered onto the sample, thus ensuring a good electrical connection between the top of the sample surface and the SEM stub. This limits charging, which can result in drifting of the sample. Moreover, gold sputtering protects the sample edges and helps reduce streaking. Figure 4.3a shows a sample after preparation, which can be used for data acquisition with the FIB/SEM system.

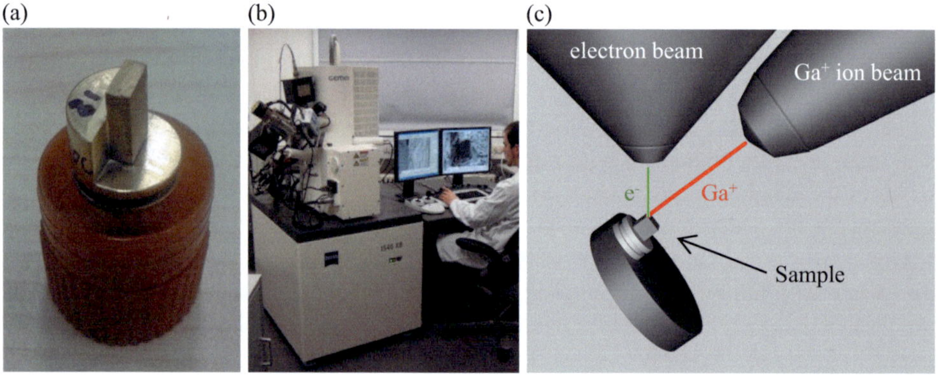

Figure 4.3.: (a) Sample after preparation mounted on a SEM stub before FIB/SEM serial sectioning, (b) SEM system used for the reconstructions and (c) schematic arrangement of the sample in the vacuum chamber of the SEM (taken from [98]).

4.3. FIB/SEM Procedure and Data Acquisition

The prepared sample is mounted on an SEM stub and placed in the SEM vacuum chamber (cf. Fig. 4.3b). It is tilted by 54°, so that the ion beam is parallel to the exposed polished cross-section that will be imaged (cf. Fig. 4.3c). The sectioning has to be conducted at a working distance of 5 mm, as this is the point where the electron and the ion beam meet.

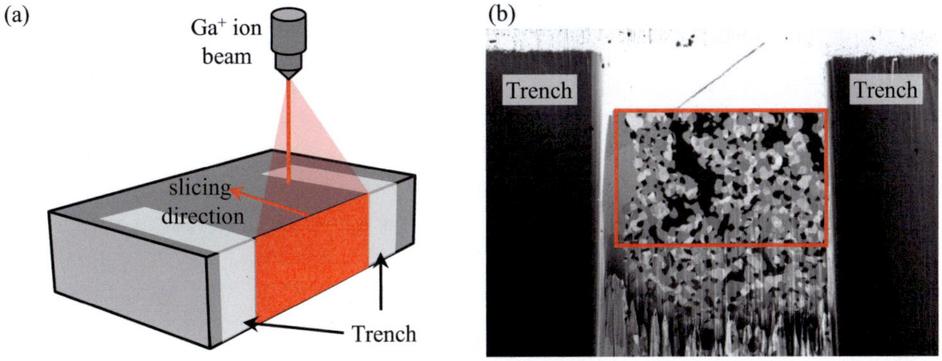

Figure 4.4.: (a) Diagram of the bulk milling technique conducted at a polished electrode edge as applied in this work: two cross-sections perpendicular to each other are polished, so that the sectioning can start directly at the sample surface (adapted from [12]). (b) An example SEM image of a prepared sample at the beginning of the sectioning. The red box indicates the maximum area which can be used for reconstruction.

As previously mentioned, most previous reconstructions in literature were done with the conventional bulk milling technique (cf. Fig. 2.9), where only one cross-section is polished and a graduated trench has to be milled before the sectioning can start. This method exacerbates many problems like re-deposition, shadowing and curtaining. To minimize these effects, Shearing *et al.*, [102] extracted the volume to be reconstructed with the FIB lift-out sample preparation technique. To eliminate the need for this time-consuming technique, but still minimize the negative effects, an advantageous preparation technique was developed within this work. This is primarily achieved prior to the sectioning by exposing the volume to be analysed. It starts by polishing two cross-sections perpendicular to each other, so that the sectioning process can start right from the edge of the sample (cf. Fig. 4.4a). Hence, it is not necessary to mill a trench in front of the ROI in order to expose it. A small trench is milled to the left and right of the analysed volume. This method minimizes shadowing effects (because the direct paths of the SE and BSE to the detectors are open), and problems with re-deposition are avoided [12, 117]. Trench-milling can be conducted at relatively high currents (5 to 20 nA) in order to obtain larger ablation rates, since the quality of the FIB polished trench surfaces is not important.

Afterwards, the sample is tilted at 90° to mill a perfect planar surface on top of the volume of interest. This provides a perfect edge between the two polished surfaces of the sample cross-sections, thus eliminating rounded edges (if introduced from polishing). In addition, the removal of different materials by the ion beam is much more uniform, giving a smooth top surface, and thus preventing streaking [98].

It is then necessary to sputter a new coating layer onto the sample, which is then returned to its original position in the vacuum chamber of the SEM. Before starting the FIB/SEM sectioning, fiducial marks can be milled into the top surface of the sample, aiding in slice alignment during image processing. However, it was found here that these marks were normally not required. Figure 4.4b shows a SEM micrograph of a prepared sample at the beginning of FIB/SEM sectioning. The red overlay indicates the maximum area which can be used for reconstruction. In order to obtain correct and meaningful results, it is important to apply the right settings for FIB/SEM sectioning. As already mentioned in Section 2.5.2, the two most important aspects regarding data accuracy are the size of the reconstructed volume and the resolution of the SEM images. Both aspects depend on the nature of the analysed structure, as well as being interdependent. For example, the resolution depends on the relevant feature sizes, since the resolution has to be sufficiently high to correctly resolve the smallest features of interest (e.g. particle sizes of the electrode structure). When reconstructing porous materials the resolution is often expressed as the number of pixels per (smallest or average) particle diameter (i.e. the ratio between the smallest particle size and the voxel resolution). To ensure a sufficient resolution in all three directions, this holds true not only for the resolution of the SEM images (resolution in x- and y-direction) but also for the slicing resolution (distance between two images, the resolution in z-direction).

However, with increasing resolution the imaged section (or *field of view*) becomes smaller and, consequently, also the volume which can be reconstructed. If the reconstructed volume is too small, the results are not representative for the whole structure and thus misleading. It is important to know the minimal size of the volume to be reconstructed which is needed to gain reliable and representative results. In continuum mechanics, the concept of a *Representative Volume Element* (RVE) is commonly used [156, 157]. Thereby, the RVE is usually regarded as a volume of heterogeneous material, which is large enough to be statistically representative for the whole structure [158]. The size of the RVE depends on different factors, like the particle size distribution or the homogeneity of the structure. The more homogeneous the structure, the smaller the RVE. Moreover, the RVE size depends on the investigated material parameter. A comprehensive discussion is given in Section 4.8, including guidelines for defining a RVE.

After defining the dimensions of the reconstructed volume, the number of cuts needed is given by the desired length of the reconstruction volume in slicing direction and the distance between the images (slice thickness). The milling time required for each slice mainly depends on the structure (mostly on the density of the materials), the length of the slices, the beam current used and the precision in focusing. A list of the most important parameters is given in Table 4.4 for two cathode reconstructions. The voxel size, reconstructed volume, detector used for the data acquisition, acceleration voltage and ion beam current are listed in Table 4.5 for all reconstructions.

Table 4.4.: Example list of properties of the reconstructed volumes and parameters used for the reconstruction of cathodes type 2_{AP} and type 2_{600}.

Parameter	Cathode Type 2_{AP}	Cathode Type 3_{1080s}
Size / number of voxel	$1150 \times 1250 \times 500$	$480 \times 550 \times 860$
Size / μm^{-3}	$40.25 \times 43.75 \times 17.5$	$12 \times 13.75 \times 21.5$
Resolution / nm	35	25
Ion beam current / nA	2	0.5
Milling time per slice / s	78	60
Acquisition time per image / s	40.1	58.8
Aperture / µm	60	30
Acceleration voltage / kV	2.0	1.2

At the beginning of this project, mostly the Everhart-Thornley detector was used for the image acquisition. Problems with charging occurred when using the in-lens detector due to the resin's poor conductivity. These problems were minimized over the progress of this project and an advanced reconstruction process using both detectors, was established (cf. Section 4.5).

The result of this FIB/SEM procedure is a stack of consecutive 2D images, where each image consists of different greyscale values, discriminated from a value of 0 (black) to a value of 255 (white). Since data acquisition takes many hours, mechanical drifts and charging of the sample can lead to shifts in the field of view. In addition, the region of interest moves slightly relative to the detector. This is due to the FIB/SEM setup, since the SEM detector is fixed while the face of interest shifts backwards with increasing penetration depth. Therefore, the images have to be aligned to each other after image collection, which is the first step in processing the consecutive images.

Table 4.5.: List of all reconstructions mentioned in this thesis, together with the most fundamental properties (ETD: Everhart-Thornley detector; IL: in-lens detector).

Cell	Voxel Size	Volume	Detector	Acc. Voltage	Ion B. Curr.
Type 1_{AP}	25-35 nm	$2517\,\mu m^{-3}$	ETD	5.0 kV	500 pA
Type 2_{AP}	35 nm	$30816\,\mu m^{-3}$	ETD & IL	2.0 kV	2 nA
Type 2_{600}	35 nm	$32049\,\mu m^{-3}$	ETD & IL	1.3 kV	2 nA
Type 2_{750}	35 nm	$21180\,\mu m^{-3}$	ETD & IL	1.3 kV	2 nA
Type 2_{900}	35 nm	$18522\,\mu m^{-3}$	ETD & IL	1.3 kV	2 nA
Type 3_{960s}	25 nm	$2906\,\mu m^{-3}$	ETD	1.3 kV	500 pA
Type 3_{1030s}	25 nm	$1519\,\mu m^{-3}$	ETD	1.2 kV	500 pA
Type 3_{1080s}	25 nm	$3548\,\mu m^{-3}$	ETD	1.2 kV	500 pA
Type 3_{1200s}	35 nm	$10022\,\mu m^{-3}$	ETD	2.0 kV	2 nA

4.4. Image Processing of the 3D FIB/SEM Data

The SEM images taken during FIB/SEM sectioning usually contain a larger area than required for reconstruction, as shown in Fig. 4.4b. This usable sector shifts upwards because the SEM detector is fixed while the face of interest shifts backwards with increasing penetration depth. This has to be taken into account during alignment. It is important to correct all shifts between the images, which can occur due to mechanical drifts and sample charging. This includes both systematic shifts (leading to a distortion of the structure) and random shifts between images (leading to a rough and, therefore, overestimated surface). Literature does not report or comment on this issue, but even a sub-pixel misalignments can add up to a significant number of pixels over a whole stack [118, 120]. This effect leads to incorrect reconstruction data from the sample microstructure and, accordingly, to imperfect microstructural parameter values (see Section 4.8). For the image alignment, the public domain image processing software ImageJ[1] with the plugin StackReg [159] is used. It calculates a correlation function for each pair of consecutive images, and determines the relative displacement of the two images at which the correlation function reaches its maximum. Before these stages the usable sector had already been cut out generously (a margin of a little under 30 pixels in all directions). This ensures that the regions in the image outside the usable sector have no or negligible influence on the correlation function. In cases where an obvious systematic drift occurs, it is advantageous to roughly correct this systematic drift first, which for this thesis was done using MATLAB [160].

If brightness gradients are present within the image data, they have to be eliminated to guarantee an adequate segmentation later on. Therefore, an algorithm developed by M. Ender [98] is used. It eliminates a gradient in one direction of the three coordinate axes. For each 2D image, the average greyscale value is calculated and then compared with the average greyscale value of the entire data set. The difference is added to the image. Afterwards, all images have the same average greyscale value. If this program is successively applied in all three directions of the data set, all gradients are balanced along the coordinate axes. Only brightness gradients caused by shadowing (typically much less pronounced along the direction axes) cannot be removed by this. However, due to the advantageous preparation described in Section 4.2, shadowing was not present within the data.

After aligning the stack of images, the images are cut to the required size. Image quality determines the number and sequence of the following filtering steps. Each set of images requires an unique filtering procedure determined by manual experimentation. An anisotropic diffusion filter was applied to all data sets analysed within this work in order to reduce noise and promote intra-region smoothing. As discussed in Section 2.6, this filter is advantageous, since it reduces the high frequency noise while retaining the particle boundaries. Where necessary a median filter was applied to some regions of the data sets. All image processing

[1] ImageJ: http://imagej.nih.gov/ij/

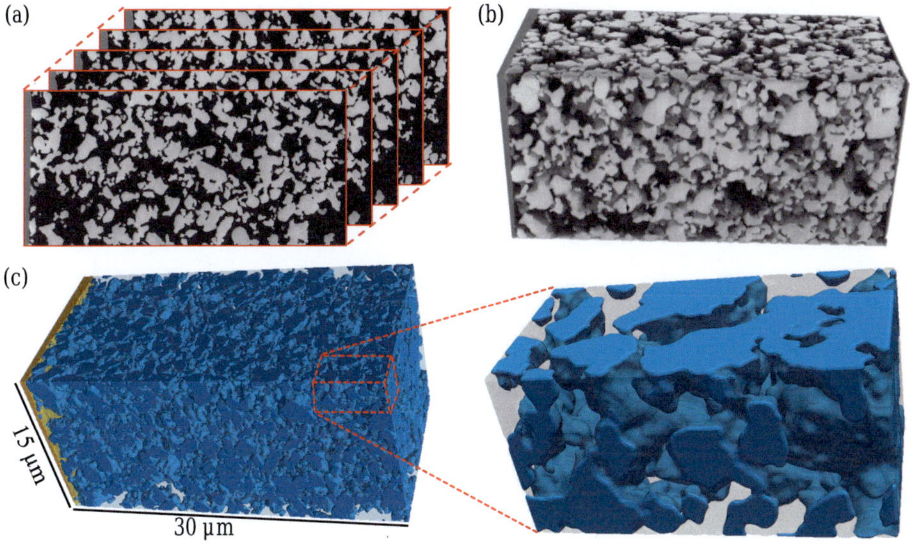

Figure 4.5.: Different steps of 3D reconstruction: (a) By stacking and aligning the 2D images in 3D space and (b) expanding the pixels in the slicing direction, a 3D reconstruction consisting of voxels is derived. (c) By assigning each voxel of the structure its corresponding phase, a 3D material distribution of the sample is obtained.

steps, except for the alignment procedure, were done with algorithms implemented in MATLAB. Figure 4.5 shows consecutive images of a porous cathode, which were stacked and aligned to obtain a 3D structure reconstruction.

4.5. Advanced Reconstruction Procedure using Two Detectors

In the beginning of this work, mostly the Everhart-Thornley detector was used solely to acquire consecutive images of the electrode microstructures, as can be seen in Table 4.5. This detector can clearly separate LSCF and pores (see e.g. Fig. 2.8a) and thereby enable the accurate reconstruction of LSCF cathodes. However, most reconstructions reported in literature were done with the in-lens detector or, more recently, also the *Energy selective Backscattered* (EsB)) detector, as they often show better contrast between materials (cf. Section 4.6). This can be advantageous when separating e.g. LSCF and GDC (cf. Fig. 2.8) or Ni and YSZ (cf. Appendix A). Next to the weaker material contrast, another drawback of using the ETD instead of the in-lens detector is the fact that the ETD is more susceptible to streaking effects (cf. Fig. 4.6). Nevertheless, the ETD also has some advantages over the other detectors. The main reason for using the ETD is that the in-lens detector can have problems during imaging when charging occurs, mainly located in isolated pores which were not infiltrated with resin (cf. Section 2.5.1). Moreover, the ETD is most suitable for the unambiguous detection of pores, as in images recorded with the ETD nearly all isolated pores appear dark like all other pores which are impregnated with epoxy. Other detectors typically do either have problems with charging in isolated pores, or electrodes from behind the pores are detected, thus making a clear segmentation difficult. This phenomenon was also observed by other groups [125] and will be shown with the example of a Ni/YSZ anode, as it most typically occurs there. Figure 4.6 shows images from both detectors focused on the same region of a Ni/YSZ anode. Please note; the contrast in the ETD acquired image is not optimized for separating YSZ and nickel, but to separate pore from the two solid phases. However, the charged areas appear very bright in SEM images acquired by in-lens detector as indicated by arrows in Fig. 4.6. This seriously interferes with automated segmentation. The SEM images revealed that these isolated pores appear more frequently within fine grained and less porous structures, e.g. in anode functional layers, which had not been analysed in reconstruction studies reported in literature.

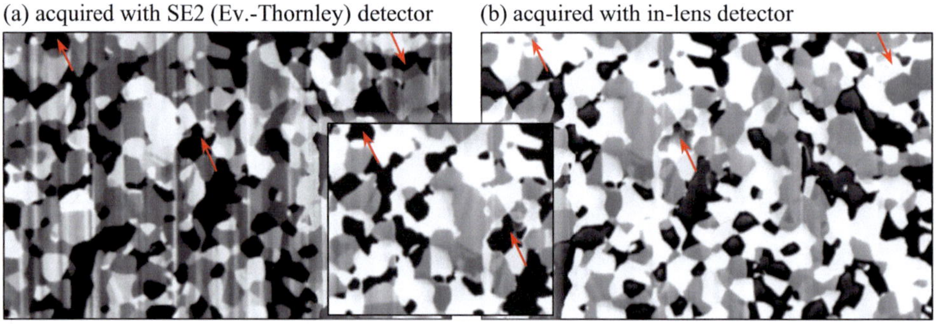

Figure 4.6.: Differences between ETD and in-lens detector (example: Ni/YSZ anode): typically in-lens images show stronger contrasts between different materials (note that the contrast in (a) is optimized for separating pore from the two solid phases, not YSZ from Ni). Also streaking appears stronger in ETD images, but they have fewer problems with charging. Advantageously, isolated and non-infiltrated pores normally appear in the correct greyscale value.

As both detectors have their advantages, the desire to combine all advantages naturally arises. The SEM system is controlled by the *SmartSEM* software. This allows the acquiring of two images simultaneously, e.g. one image taken with the ETD and the other using the in-lens detector. The image acquisition time is unaffected. The idea is to then optimize the contrast between the pores and the solid phases of the ETD images for subsequent pore identification. In the images acquired with the in-lens detector, the contrast between the solid phases is optimized. Thus both datasets can be combined to obtain an accurate identification of all phases which are present in the electrode.

As discussed in the previous section, the consecutive images must be aligned and the usable sector cut out of the larger image. Since there are now two stacks of consecutive images (one acquired by ETD and the other by in-lens detector) it is necessary to align and crop both stacks identically. Otherwise the information from both detectors can not be combined. Therefore, first one of the stacks is cropped and aligned using the ImageJ software (see Section 4.4). During the alignment process each shift, change and cut is recorded. This information is then used to identically align the second stack of images in MATLAB. Naturally, the cutting must also be performed identically. The enlarged section in the middle of Fig. 4.6 shows the in-lense image (b) after it was combined with the pores identified using the ETD image (a).

4.6. Segmentation

To obtain a 3D material distribution of the structure, as for example shown in Fig. 4.5c, the greyscale images must be segmented. More precisely, the entire greyscale image, voxel by voxel, is unambiguously assigned to one of the structural phases. As mentioned before, this segmentation process is the most important step in image processing and potentially a large source of error. The simplest and most widespread method for the segmentation of pixel- and voxel-based data is to use a threshold value [99, 161]: all voxels having a greyscale value within a certain range can be partitioned into the same segment. But despite its apparent simplicity, thresholding is a powerful, yet challenging, method (cf. Section 4.6.1).

Threshold segmentation requires excellent quality image data. The image data histogram should have clearly separated peaks, where each peak corresponds to a single phase. The true threshold value is located somewhere in the "valley" between two peaks (cf. Fig. 2.13). If the peaks are not clearly separated, thresholding will lead to segmentation errors, resulting in noise on the segmented images. Although these errors can later be partially corrected with morphological filters, it is possible (and easier) to have already counteracted these errors during the segmentation process. For instance, information on a pixel's surroundings can be used to decide which phase the pixel belongs to. This can be done in an iterative process, e.g. by looking at already assigned neighboring pixels, or by using the local gradient around a pixel. Examples of such segmentation methods are *watershad* and *region growing* methods [99]. Generally, the appropriate method for segmentation is strongly dependant on the nature and quality of the images. There is no single, universally applicable method for the correct segmentation of images. Accordingly, a method for the adequate segmentation of the image data of our porous electrodes was developed within this work. This method belongs to the region growing class of segmentation methods, and will be presented in the following subsections for the segmentation of two and three phase structures.

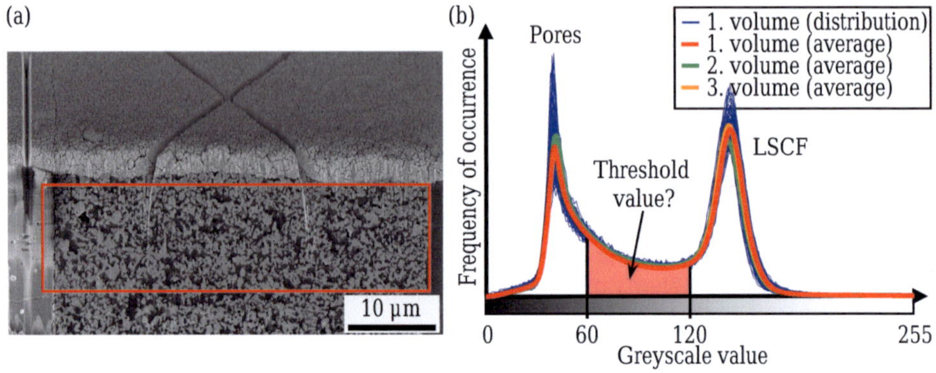

Figure 4.7.: (a) SEM image of a porous LSCF cathode from the first reconstruction performed during this thesis. (b) Corresponding histogram of 213 images (thin blue lines) together with the averaged histograms of all these images (red line), as well as the average histogram of two other volumes (green and orange) acquired from the same sample.

4.6.1. Segmentation via Thresholding

As already discussed, the most common method of segmentation is thresholding, and it requires excellent image quality. To check if an automated segmentation with a threshold value is expedient, a shape analysis of the histogram can be helpful. To illustrate this process, the first reconstruction taken within the course of this thesis (cathode type 1) is used as an example. Over 600 images of a porous LSCF cathode were acquired and three non-overlapping volumes were reconstructed from this data. Each volume uses 213 images (cf. [117] for more details). Figure 4.7 shows one of the images, together with the histogram of all 213 images used for the reconstruction of the first volume (thin blue lines). Moreover, the average histogram of the first volume is displayed in red, alongside the histogram for two other volumes (in green and orange).

An admissible precondition for global threshold segmentation of structures consisting of two phases is a bimodal distribution of the greyscales values, which is obviously fulfilled here (cf. Fig. 4.7). Poor image contrast leads to a histogram where peaks tend to overlap, which makes it difficult to separate pores from electrode material. Image shadowing artefacts are even more serious (cf. Section 4.2), leading to images appearing lighter at one side and darker at the other. Because many images are needed (n.b. between 200 and 800 images per reconstruction within this work), all of them should have the same contrast and the absence of a brightness gradient between the consecutive images is a further prerequisite. This can be controlled by comparing the individual histograms of the subsequent images. If the peaks of different images are not shifted relative to each other, it can be concluded that there is no brightness gradient and no image shadowing. It can be recognized from the greyscale frequency distribution of the 213 individual images (cf. Fig. 4.7b) that the distribution of the two peaks is well aligned. Consequently, brightness gradients or shadowing are of no concern and segmentation with a global threshold value is permissible. Otherwise, a prior correction of the brightness gradients (as discussed in Section 4.4), or a local threshold segmentation method (as described e.g. in [98]) would be necessary.

Since the preconditions for segmentation by global thresholding are fulfilled, the question arises: which greyscale value represents the true threshold? In the histogram of SEM images the peaks are normally imperfectly separated, due to various reasons. The most important reason for this is the limited resolution of SEM images, which is why the boundaries between phases are never distinct. Roughly speaking, between a phase that appears light and a phase that appears dark, there will normally be an interlayer (at least one pixel wide) with an intermediate greyscale value (if only because the area represented by this pixel contains, most probably, a part of both phases). Many other reasons may additionally intensify this effect (e.g. an imperfectly focused electron beam).

In Fig. 4.7b the peaks are poorly separated, which makes it hard to identify the true threshold value. Figure 4.8 shows two examples of a poorly chosen threshold: if the corresponding greyscale value is too low (with a threshold value of 60 in Fig. 4.8a), a significant part of the pore volume is attributed to electrode material. If the corresponding greyscale value is too high (with a threshold value of 120 in Fig. 4.8b), electrode material may be attributed to the pore phase. Obviously, the resulting porosity/material fraction would be inaccurate in both cases. How much the parameters change with the chosen threshold value is investigated in Section 4.8.3.

(a) Threshold value = 60 (b) Threshold value = 120

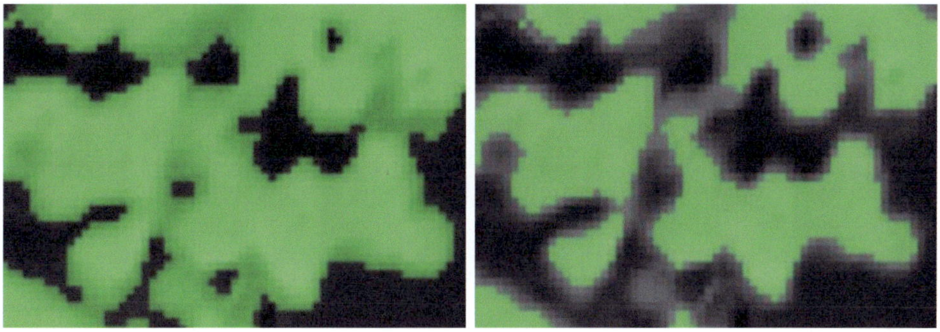

Figure 4.8.: Examples for a poorly chosen threshold (segmentation of LSCF in green): (a) If the threshold is chosen too low, parts of pores are attributed erroneously to electrode material (dark green regions), while (b) if the threshold is chosen too high, parts of electrode material are attributed erroneously to pores (light grey regions).

Locating the true threshold value is one of the most important reconstruction tasks. It can be roughly set by interpreting the histogram shape. Given a bimodal distribution, the minimum or "valley-floor" between the peaks is the obvious choice for the threshold. Even in the case of two peaks that are not clearly resolvable (like in the example) several algorithms are available. Most of them are based on statistical interpretation techniques and try to automatically locate the optimal threshold value.

Algorithm for threshold determination

Since no clear minimum could be observed between the histogram peaks (cf. Fig. 4.7), several methods were evaluated for the automatic identification of the optimal threshold value for this data set. A widely used method is the Ptile- or Median-algorithm [162]. This algorithm works well if the volume fraction is already known. This method is obviously unsuitable here, because the volume fractions are unknown and their determination one of the goals of this analysis. The Mean-algorithm is a very intuitive method of finding the appropriate threshold; calculating the mean value of all greyscale values. It determined a threshold value of 100 for this data set and it is discussed alongside many other methods in Ref. [163].

Otsu's method [164] is more advanced, and is one of the most qualified methods for image data consisting of two phases (here: pore and LSCF). The two phases are considered as two clusters of greyscale values C_1 and C_2, which are defined as

$$C_1 = \sum_{i=0}^{T} p(i) \text{ and } C_2 = \sum_{i=T+1}^{255} p(i), \tag{4.1}$$

where $p(i)$ is the amount of pixels at greyscale value i, defined as the number of pixels n_i at level i divided by the total number of pixels N. The algorithm settles both clusters as tightly as possible. As the threshold value is adjusted, the spread of one cluster increases, while the other decreases. The goal is to then minimize the combined spreads or equivalently maximize the variance between the clusters. First, the variances of the two clusters of intensity values separated by an intensity threshold t are calculated by using the mean values μ_1 and μ_2 of their greyscales

$$\sigma_1^2 = \sum_{i=0}^{t} (i-\mu_1)^2 p(i) \text{ and } \sigma_2^2 = \sum_{i=t+1}^{255} (i-\mu_2)^2 p(i). \tag{4.2}$$

Second, the variance between the clusters is calculated by using the mean value of all greyscale values μ

$$\sigma_{Bi}^2 = C_1(\mu_1-\mu)^2 + C_1(\mu_2-\mu)^2. \tag{4.3}$$

Now, the function

$$\frac{\sigma_{Bi}^2}{\sigma_t^2} \rightarrow \max. \qquad (4.4)$$

is calculated for each intensity $i = 0, \cdots, I-1$, where $\sigma_t^2 = C_1\sigma_1^2 + C_2\sigma_2^2$ is the total variance. The intensity which maximizes function (4.4), is said to be the optimal threshold.

The threshold values resulting from the Otsu-algorithm (global threshold value 97) and the Mean-algorithm (global threshold value 100) are rather close to each other, thus the calculated porosity in this example only differs by about 1.2 % (absolute), see Section 4.8. Nevertheless, the Otsu-algorithm is favoured here, since this algorithm makes use of more information by calculating the variance of the two phases. Interestingly, the same threshold value of 97 was extracted by "intuitive" inspection of the histogram. Also, Otsu's method showed the best results for other data sets. But irrespective of which method used, a careful inspection of the resulted segmentation is always advisable.

However, if the greyscale values of the different phases can not be unambiguously segmented via thresholding, other methods must be used. Therefore, an algorithm belonging to the *region growing* methods was developed within this thesis, which will be presented in the next subsection. Although this algorithm is introduced for the segmentation of three phases, it can be easily applied for more or fewer.

4.6.2. Segmentation via Region Growing

Finding a good contrast between the different phases in SOFC electrodes can be highly complicated. This holds especially true for Ni/YSZ anodes [96], which is why the introduction of this algorithm uses the example of Ni/YSZ anodes. Looking at three-phase electrodes (Ni, YSZ and pores) helps to explain some particularities occurring during multiple phase segmentation. Nevertheless, the principle of the method is the same for segmenting two-phase electrodes (e.g. LSCF and pores).

Poor image contrast can cause histogram peaks to overlap, and then thresholding leads to segmentation errors. Figure 4.9a shows the histogram of an anode imaged with an ETD. It indicates that the pore phase is easily separable by a threshold value, but the greyscale values of Ni and YSZ overlap. Applying a threshold segmentation would lead to parameter distortions; to e.g surface area and triple-phase boundary density. This is illustrated in Fig. 4.9b which shows an image after threshold segmentation. The areas marked by an oval contain many incorrectly assigned pixels, either YSZ (grey) in a large Ni (white) matrix or vice versa.

However, regardless of data quality, segmentation of three phases via thresholding will never be perfect. This is depicted in Fig. 4.10, where an anode image acquired by an in-lens detector with a very good contrast between the phases is shown. The enlarged section in (b)

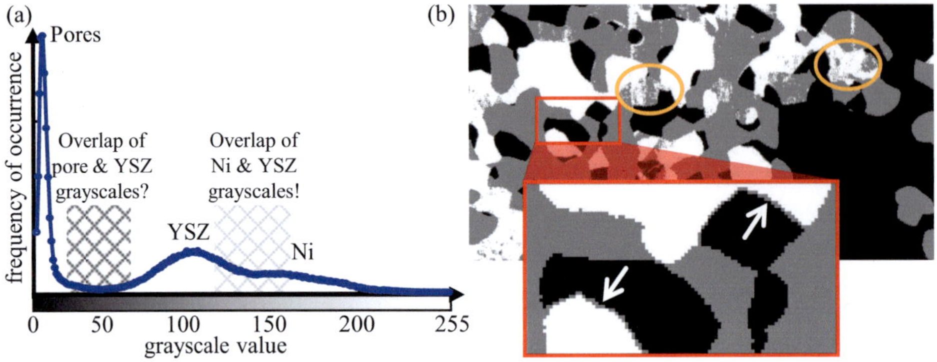

Figure 4.9.: Problems with threshold segmentation: (a) Histogram for one anode image, acquired by ETD. The greyscales for the pore phase (dark) are separable from the two solid phases YSZ and Ni, but the brightness values of Ni and YSZ partly overlap. (b) image after threshold segmentation: Some areas (e.g. as marked by an oval) contain many incorrectly assigned pixels, either YSZ (grey) in a large Ni (white) matrix or vice versa. The magnification shows that even areas which seems to be correctly segmented show a thin layer of YSZ generated erroneously at the borderline between pore and Ni phase.

shows that at the interface between areas with dark greyscale values and areas with bright greyscale values there is always a natural transition of intermediate greyscale values. By applying threshold segmentation, these grey interlayers will be erroneously assigned to third phase, with the intermediate greyscale values, as can be seen in the enlarged sections of Figs. 4.9b and 4.10c. As a consequence, triple-phase boundaries are undetectable and surface areas are either overestimated (phase with intermediate greyscale values) or underestimated (surface between dark and bright phases). These problems might be solved by a combination of thresholding and morphological operations [138], but this task has to be performed very carefully without changing the shape of the remaining microstructure (cf. Fig. 2.14).

To overcome these problems and achieve an accurate data segmentation, a segmentation algorithm inspired by Jørgensen at el. [138] was developed, which belongs to the region-growing image segmentation methods [165]. The first version of this algorithm was developed in the bachelor thesis of Ingo Rotscholl [166]. That thesis was supervised in the context of this thesis, and the algorithm has been under continual development ever since. Details of its first implementation can therefore be found in [166].

The basic idea of region-growing is straightforward and will be explained in the context of an anode dataset: starting from a set of voxels which have a high probability of belonging to a specific phase (referred to as "seeds"), this subset of voxels is grown according to certain requirements, until all voxels are assigned. Thereby, the algorithm simultaneously performs the segmentation of all three phases in 3D in an iterative growing process.

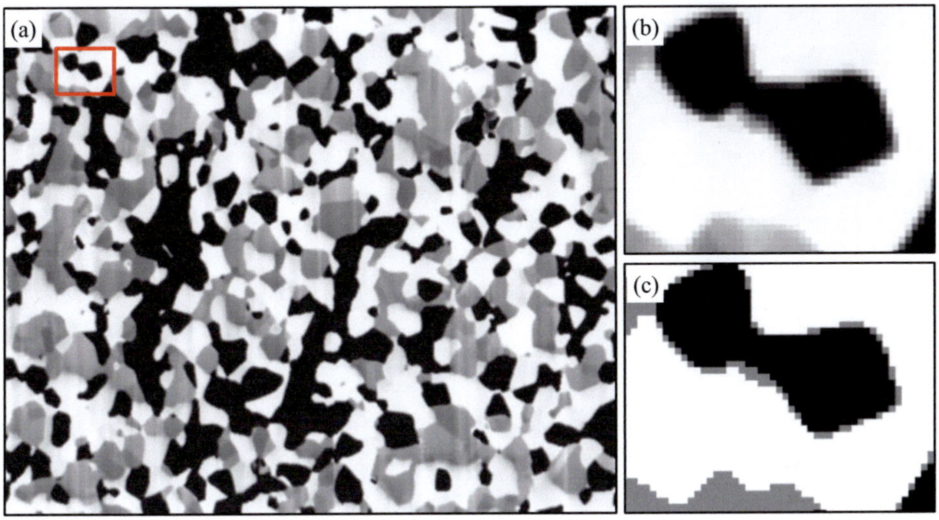

Figure 4.10.: Segmentation error "grey interlayer": (a) SEM image acquired by in-lens detector of an anode after image processing. (b) The magnification indicates the grey interlayer, naturally occurring between dark (here: pore) and light (here: YSZ) areas. (c) Thresholding assigns this grey interlayer erroneously to the intermediate grey phase. As a consequence, false triple-phase boundaries are created and surface areas are either overestimated (grey phase) or underestimated (surface between black and white phase).

However, a preprocessing step is required. The boundaries between the phases (which can be interpreted as edges inside the data set) are detected with an edge detection filter and all voxels belonging to an edge are labeled as "edge". This information is stored in a so-called "edge-map", which is settled here using the Canny-algorithm [167].

Then, all voxels that almost certainly belong to a specific phase are set as a starting point ("seed"). These seeds can be identified by a careful inspection of the histogram. In this instance the seeds for the pore phase would be a small subset of voxels with a very low intensity (e.g. greyscale values < 25; cf. Fig. 4.11). In contrast, the seeds for the Ni phase would be high intensity voxels that are especially different to those of the YSZ phase (e.g. greyscale values > 170). In addition, the seeds for the YSZ phase should be of a higher intensity than the pore phase, but low enough so as to not belong to the Ni phase (e.g. greyscale values in-between 70 and 100 in Fig. 4.11). Moreover, the absence of a greyscale gradient in the neighbourhood of the Ni phase seed should be ascertained, as this indicates the gradient between the dark pore phase and the bright YSZ phase.

For an automated identification of the seeds, the histogram shown in Fig. 4.11 is first divided into three sections by two classical segmentation methods. For the structures investigated here, Otsu's method [164] is considered the most suitable for separation of pore and YSZ, and Maximum Entropy [168] for the separation of YSZ and nickel. As a consequence, threshold calculations have to be performed separately for each single image, since the consecutive images could have a brightness gradient. These gradients do not appear with the same distinctness for the different phases, as the pore phase is often unaffected while YSZ can show pronounced gradients. To overcome this constraint, the obtained threshold values were linearly interpolated. Advantageously, the resulting dataset did not contain any discontinuities. The correctness of our segmentation procedure was confirmed by a subsequent EDX analysis of the last image [166].

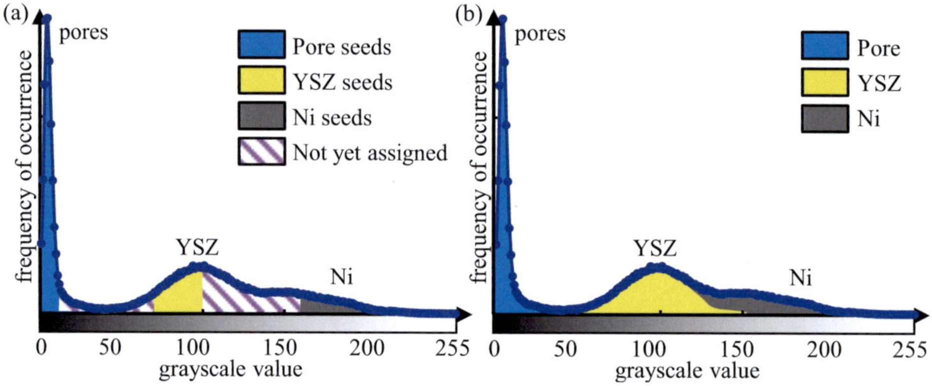

Figure 4.11.: *Region-growing* algorithm: (a) schematic histogram at the start of the segmentation procedure, with greyscale values assigned to seeds of the pore phase (blue), YSZ phase (yellow), and Ni phase (grey). (b) Schematic histogram at the end of the segmentation procedure showing the assignment to all phases.

The second step is the iterative growing process. Spreading out from the seeds, all "not yet assigned" neighbouring voxels which share a surface with one of the seed-voxels are checked to determine whether they belong to the same phase. If a voxel is assigned the same phase, this voxel is added to the seed and the next iteration step begins. As a general rule, growing across edges is not allowed, which is controlled by utilizing the edge-map. However, phase growing requires additional rules which cooperatively identify where the growth of a specific phase ends. Altogether, four decision criteria are established:

1. Growing across edges is not allowed, which is controlled by using the *edge-map*.
2. Probability that the greyscale value of the approached voxel still belongs to the phase of the seed - uncertainties are highest on or near the threshold values.
3. Difference in greyscale values of two neighbouring voxels (between seed-voxel and considered voxel): the smaller the difference, the greater the probability that they belong to the same phase.

4. Number of neighbouring voxels already assigned to one of the phases: the more neighbouring voxels belong to phase i, the more likely it is that this voxel also belongs to phase i.

These criteria can be weighted according to the requirements and the image data quality. For example: criterion (2) should have the highest weight if the brightness values of all phases are clearly separable, while criterion (4) should have a higher weighting if the brightness values of the phases are similar. Additionally, the weight of criteria (1) to (4) can vary during the segmentation process of a large image data set. Labelling voxels with a greyscale value on or near one of the two threshold values is most critical, as they have the same probability of belonging to both phases. In these critical areas, all four criteria will be taken into account. In contrast, the first two criteria are sufficient when voxels have a greyscale value which is clearly different to the thresholds. As mentioned before, all phases grow simultaneously, and the iterative process of segmentation is completed once all voxels of the FIB/SEM data set are assigned to one of the three phases.

The third step is an optional morphological operation, and is performed after the growing process. In a few cases this step is necessary, e.g. if the grey interlayer is erroneously assigned to the intermediate greyscale phase (indicated by the arrows in the enlarged section in Fig. 4.9). By a morphological operation, the phases shrink together and then expand, removing these slim regions. The removed regions have to be reassigned, where the reassignment re-applies the four criteria of the growing process.

As a strong point, our region growing algorithm considers both, information on greyscale values and geometrical information like boundaries and gradients between two adjacent phases. This approach results in a more reliable separation result, especially for images showing low contrast between the phases. For example, the watershed method results in an overestimation of the Ni phase, which is amplified for noise-affected image data. Another advantage of this algorithm is that the segmentation is completed in 3D. Thus, if e.g. looking at the neighbourhood of a voxel, a third of the voxels come from the image in front of the voxel to be assigned, a third come from the same image, and a third come from the image behind the voxel in question. This can significantly improve the consistency of data along the z-dimension (slicing direction).

As a weak point, our region growing algorithm requires the definition of reliable and certain seeds at the beginning of the region growing process. If one only uses greyscale values to define the seeds, there is a danger that the grey blurred interlayers between dark and light particles can become seeds of the intermediate phase. To avoid this, a later version of the algorithm was enhanced to identify these interlayers. Image gradients were inspected, and seeds from high gradient areas were disallowed (on or near edges). Such problems can also be avoided by applying the morphological operation *erode* to the seeds. Another difficulty, shared with almost all other segmentation methods, is the appearance of twin boundaries in the Ni phase, which can often be seen using low energy imaging. The Canny algorithm will detect an edge inside the nickel grain, which may affect the segmentation result. With our region growing algorithm, two cases are probable: (i) a correct result if Ni seeds are detected on both sides of the twin boundary, (ii) an incorrect result if a Ni seed and a pore or YSZ seed are detected. In this case, the phase assignment has to be corrected manually.

This *region-growing* algorithm was first developed for Ni/YSZ anodes consisting of three phases, but within this thesis it is has also been used for the segmentation of LSCF cathodes consisting of only two phases. Moreover, this algorithm has the potential to segment structures consisting of more than three phases.

4.7. Parameter Calculation: Quantification of Electrode Microstructures

After segmentation, the 3D data set consists of i phases where all voxels of the same phase are assigned to the same value. Relevant parameters that characterise a microstructure can be calculated based on these segmented data. This section presents numerous algorithms developed for the calculation of these parameters and discusses some important aspects for the accuracy of the methods. Although the different methods have a great influence on the calculated values, there are no established methods of identifying these parameters from 3D image data. An extensive discussion of accuracy aspects regarding the parameter identification is given in Section 4.8. However, the calculated parameters can be used to discuss electrode microstructural differences between different samples. This is presented in Section 4.9. Moreover, the parameters can also be used to parametrize (partly-) homogenized microstructure models (e.g. Refs. [60] or [169]) to predict the electrode performance (cf. Section 5.6).

4.7.1. Phase Connectivity or Percolation

The connectivity of phases (also called percolation or inter-connectivity in literature) is an important characteristic, especially when investigating the interplay between microstructure and electrode performance. Only sufficiently connected parts of the microstructure can contribute to the cell reactions, isolated parts naturally cannot. However, studies in literature have used different definitions to define connected particles or pores [170], which should be considered when comparing the results of different groups. Moreover, in almost all cases only a small part of the full electrode height was reconstructed. Therefore, connected parts will in most cases be overestimated even if for large volumes the uncertainty of the results should be small. In this work the connectivity of the different phases is calculated using MATLAB [160]. By using the following definition, all voxel clusters are labelled as:

1. *isolated* within the reconstructed volume (clusters that lack contact to the connected phase of the reconstruction),
2. *unknown*, i.e. they appear isolated but intersect one of the outer faces of the volume so that the true connectivity cannot be established or they are connected by only an edge or vertex,
3. *connected*, if connected to either the reconstruction's electrolyte face or current collector/gas channel face or to both faces, depending on whether the cluster is ionic, electronic or mixed conducting or porosity.

In other words, LSCF voxels are considered connected when they percolate with the electrolyte and the current collector and pores are consider connected if they percolate with the gas channel. Note that only neighbouring voxels of the same phase, sharing a common face, were considered to be connected. However, sometimes in literature voxels are also considered to be connected if they share a common edge or even only a vertex. Nevertheless, the connectivity analysis results should be similar, irrespective of the definition of connectivity between the voxels, which is the case for our reconstructions. If not, a significant number of particles would only be connected via voxel-vertices and not by voxel-faces as well. This in turn indicates that the image resolution should be increased to more accurately resolve the connections between particles. Now all voxels are labelled with their connectivity status, this information can be combined with the calculation of other parameters like volume fraction and surface area. This will identify potentially active and isolated volume fractions and surface areas.

4.7.2. Material Fraction

The most fundamental characteristic of porous electrodes is the volume fraction V_i of each phase. These values are calculated by dividing the number of voxels, N_i, assigned to the specific phase i by the total number of voxels N

$$V_\mathrm{i} = \frac{N_\mathrm{i}}{\sum_i N_\mathrm{i}} = \frac{N_\mathrm{i}}{N}. \tag{4.5}$$

Naturally, erroneous segmentation directly affects the calculated volume fractions (cf. Section 4.8.3). To which degree must be individually examined for each data set, as this depends on the data characteristics and the method used for segmentation.

A reconstructed volume which is not large enough will lead to a statistical error because the results do not represent the whole structure. This means that even a small change in the observed volume can noticeably change the results. To estimate the magnitude of this effect for a particular case, a statistical analysis has to be performed.

A further source of error is insufficient resolution of the microstructure. This can lead to an over- or underestimation of the volume fractions. This effect can be investigated by using artificial microstructures and it will be discussed together with the influence of resolution on the surface area in Section 4.8.1 (cf. Fig. 4.17).

4.7.3. Surface Area

Surface area is one of the most important parameters of porous electrodes, as their size is correlated to the electrochemical reactions in the electrodes (cf. Section 2.4). In order to compare the surface areas of various samples, or reconstructions which differ in volume size, the surface area is expressed as a (volume-specific) surface area density. This means that the total surface area A, of a specific phase is normalized to the total reconstructed volume. The volume-specific surface area will be denoted as a_i. It is important to *not* consider the surface areas at the face of the reconstructed volume. However, it should be mentioned that in some publications (e.g. Ref. [125]) the surface area of a specific phase is normalized by the volume of that phase and not, as here, by the total volume. This should be considered when comparing values.

The most intuitive way of calculating the surface area A is to count all voxel-faces between two different phases and multiply the number $N_{\text{interface}}$ by the area A_{vf} of one voxel-face

$$A = N_{\text{interface}} \cdot A_{\text{vf}}. \tag{4.6}$$

While this is useful, the surface of rounded particles will always be overestimated if the particles are approximated by voxel, which can not be prevented by a higher resolution. The overestimation reaches its maximum with the case of a perfect sphere where the surface area can be calculated by $4\pi r^2$ (where r is the radius of the sphere). For a sphere approximated by voxels the surface area tends towards $6\pi r^2$ (six times the area of a circle with the same radius) with increasing resolution. This can be easily understood if looking at the sphere from each of the six directions parallel to the coordinate axes: The visible voxel-surfaces will outline a circle [98]. This means that in the limit the surface area will be overestimated by 50 % (cf. Fig. 4.17) [104].

It is therefore necessary to calculate the surface area with another method. An appropriate method is the *marching cubes* (MC) method [171], which extracts a polygonal mesh of an isosurface from 3D voxel data. The *marching cube* consists of 8 neighbouring voxels and the surface area is approximated by triangles in dependence of the material distribution of these 8 voxels. This is done by looking at whether or not two neighbouring corners of the cube (connected by an edge) belong to different phases. If they belong to different phases the midpoint of the edge is tagged and constitutes a corner of a triangle which builds the surface of the structure.

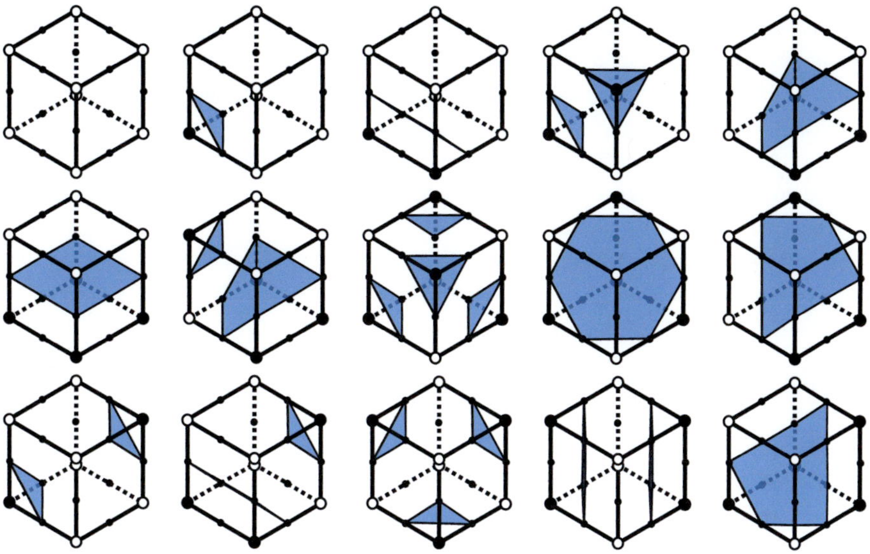

Figure 4.12.: Calculating the surface area with the *marching cubes* algorithm: the algorithm reduces the 2^8 possible cases of material distribution to 15 unique cases by exploiting symmetries and also signal changes (adapted from [98]).

This is done with any eight neighbouring voxels of the structure. In the case of two phases there are $2^8 = 256$ possibilities of distributing the 2 phases in the 8 voxels. The algorithm reduces the 256 possible cases to 15 unique cases by exploiting symmetries and signal changes (cases with interchanged phases belong to the same unique case), which are illustrated in Fig. 4.12. By counting the number of each of these unique cases within a reconstructed volume and multiplying this with the surface area of the triangles of the associated case, the surface area of the structure is calculated. The surface area density is obtained by dividing this area by the volume. It should be considered at this point, that the volume in each direction with N_i voxels contains only $N_i - 1$ cubes. Hence, the reference volume V_{ref} is smaller than the total volume:

$$V_{\mathrm{ref}} = (N_x - 1) \cdot (N_y - 1) \cdot (N_z - 1). \tag{4.7}$$

Although the surface area can be approximated much more precisely with the marching cubes algorithm compared to approximating with voxel-mesh, the surface area will still be overestimated [104]. Here, too, the overestimation reaches its maximum for spheres, but with only about 7 % in the limit. Naturally, as for all other parameters, the calculated values depend on factors like resolution and volume size, as will be discussed in Section 4.8.

It is important to notice that for irregularly shaped particles (as present in SOFC electrodes) the deviation from the actual surface area will be smaller than in the "worst-case" example of spheres. How much by depends on the structure, and especially on the shape of the particles and is hard to assess. Thus it is not generally possible to correct the results. When interpreting the calculated surface areas of a real electrode structure, it must be noted that the surface of a poorly resolved phasc (phase with small particles) may be underestimated. Conversely, the surface of phases with larger particles may be overestimated due to the method and the better resolution of these larger particles [98]. When comparing surface areas calculated by different groups, the method used for calculation must be considered.

4.7.4. Tortuosity

The tortuosity τ is a very characteristic property of porous microstructures as it provides a measure for how twisty a transport path is. It is of particular importance when looking at the transport of different species inside the phases (i.e. gas transport inside the pores or the ionic/electronic transport in the material phases). Together with the volume fractions V_i, the tortuosity connects the intrinsic bulk conductivity σ_{bulk} (or diffusivity) with the effective conductivity σ_{eff} of the porous structure

$$\sigma_{\text{eff}} = \frac{V_i}{\tau} \cdot \sigma_{\text{bulk}}. \tag{4.8}$$

For straight pathways (e.g. from top to bottom in Fig. 4.13a) the tortuosity is equal to 1. In the case of there being no continuous pathway in the investigated transport direction, τ would be infinite (e.g. from left to right in Fig. 4.13a). However, in literature the tortuosity is the parameter that causes the most confusion. This is mainly because different definitions exist for the tortousity and there are also different ways to calculate it. A discussion of the different definitions can be found in [172]. Since the transport of a specific species is normally only considered to take place in one of the phases, only one conductive phase is assumed in the following.

It is important to know that the tortuosity of porous structures can be interpreted in two ways. Either in terms of an effective medium [173] or geometrically, where it is a measure for the elongation of a transport path due to the twisty structure of the pores. Geometrical approaches ([134, 172, 174]) define the tortuosity e.g. as the ratio of the diffusion length or "effective average path length" L_{eff} to the geometric length L (for example the electrode thickness), i.e.

$$\tau_{\text{geom}} = \frac{L_{\text{eff}}}{L}. \tag{4.9}$$

Please note: if the prolongation of the transport path is used to define the tortuosity (as given in Eq. (4.9)), then Eq. (4.8) must be written with τ^2 instead of τ.

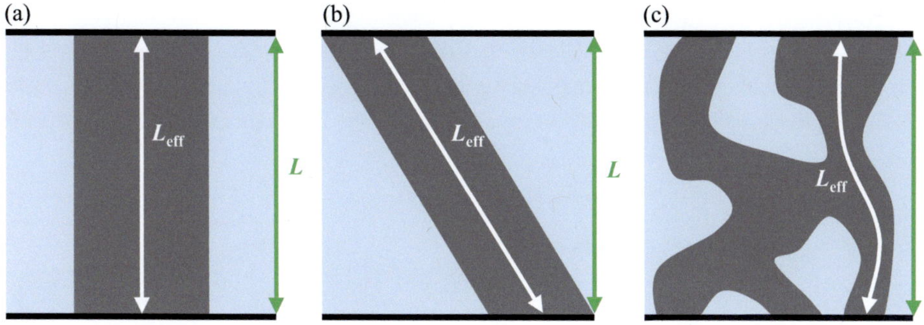

Figure 4.13.: 2D illustration of the geometrical tortuosity τ_{geom} showing conductive pathways in an isolated matrix: (a) For straight pathways the tortuosity is equal to one, in the case of non-continuous pathway it is infinite. (b) The longer the pathway, the higher the tortuosity value. (c) In the case of connected pathways, calculating the geometrical tortuosity is not readily possible.

If comparing two pathways of different length but with the same material fractions (e.g. Fig. 4.13 (a) and (b)), the effective conductivity of the longer pathway will be lower. One reason is the extension of the transport path by a factor of L_{eff}/L, but one should also take into account the reduction of the cross-section of the pathway (which is necessary to obtain the same material fractions) by the same factor L_{eff}/L. In this case the (geometrical) tortuosity would be [98]:

$$\tau_{\text{geom}} = (\frac{L_{\text{eff}}}{L})^2. \tag{4.10}$$

In the case of non-connected pathways of constant diameter throughout the structure, the geometrical tortuosity can be calculated. For more realistic structures with interconnected pathways of varying diameters, an exact geometric calculation of the tortuosity is not possible. This approach does not account for pore constrictions which would naturally influence the effective conductivity. It is nevertheless used for an approximate calculation of the tortuosity, where the effective length is approximated by the average length of random paths through the structure (e.g. calculated by random walk [133]). To account for the varying diameters of the pathways, an additional correction factor called *constrictivity* is occasionally used [175]. However, an exact calculation of the constrictivity for real structures is not possible [98].

To avoid all these problems, the effective media definition as e.g. in [86, 98, 117] using Eq. (4.8) is used here to calculate the tortuosity. Thereby it should be noted that since the definition of tortuosity is inconsistent across literature, the definition of Eq. (4.8) as used here is sometimes denoted as *tortuosity factor*; in which case the square root of it is then called tortuosity. For calculating the tortuosity, a finite element method (FEM) model was implemented as described in [120]. The mesh for the calculation was created by taking the

3D reconstruction data after segmentation and converting each voxel of the analysed phase into an 8-node cubic element. This reflects the natural spatial resolution of the datasets. Thereby, only voxels which are connected by faces (and not only by edges or vertices) were added to the mesh, in order to avoid singularities. A schematic design of the model is given in Fig. 4.14.

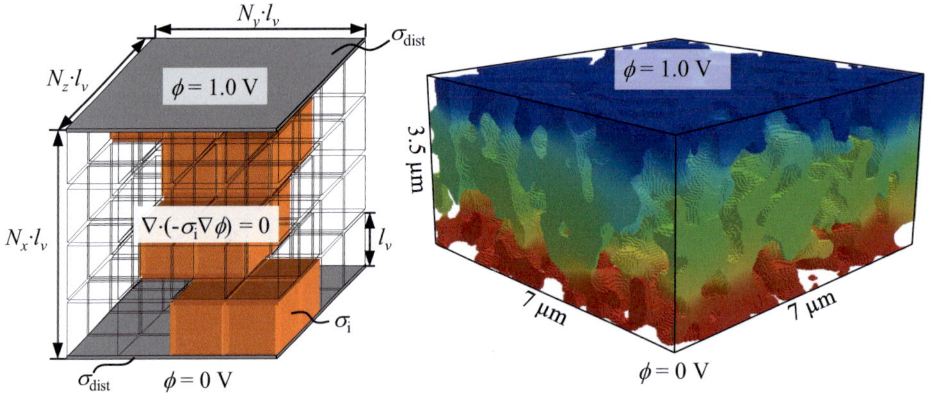

Figure 4.14.: Schematic design of the finite element method model (left) which was applied for the calculation of the tortuosity and an example of the potential distribution ϕ inside a part of a reconstructed microstructure (right).

The tortuosity is determined in transport direction from the electrolyte/cathode-interface to the current collector for one of the phases i. On top and bottom of the reconstructed volume two thin layers with a very high electrical conductivity σ_{dist} are applied to distribute and collect the current to the porous electrode structure (cf. Fig. 4.14). A potential difference η_{Model} between these distributing layers is predefined as 1.0 V by applying a potential $\phi = 1.0\,\text{V}$ on the upper layer and $\phi = 0\,\text{V}$ on the bottom layer. At the remaining four external faces of the reconstructed volume, as well as on the interfaces to the other (not considered) phases, a zero flux boundary condition is applied. Inside the considered phase, the diffusion transport equation

$$\nabla \cdot (-\sigma_{\text{bulk,i}} \nabla \phi) = 0 \tag{4.11}$$

is solved to determine the current I that circulates through the structure. Thereby the intrinsic conductivity of the material $\sigma_{\text{bulk,i}}$ is predefined. Afterwards the resistance R of the structure is calculated by using Ohm's law. Hence the effective conductivity is given as

$$\sigma_{\text{eff}} = \frac{I}{\eta_{\text{Model}}} \frac{N_x}{N_y N_z l_v}, \tag{4.12}$$

where N_x, N_y and N_z are the number of voxels in x- y- and z-direction and l_v is the edge length of the cubic voxels. Finally the tortuosity (τ_i) can be calculated by the formula

$$\tau_i = V_i \frac{\sigma_{\text{bulk,i}}}{\sigma_{\text{eff}}}, \tag{4.13}$$

with the previously determined volume fraction V_i.

Calculations based on such complex structures, with such a high resolution, require a lot of computational effort. Therefore a 3D FEM model was developed in close collaboration with T. Carraro (Institute for Applied Mathematics, Heidelberg University). It is implemented in the in-house software ParCell3D based on the C++ FEM library deal.II [176] and solves the model equations by means of high performance computing (HPC) techniques. More details of this model can be found in Refs. [119, 177] and will also be provided in Chapter 5.

Within this work a MATLAB implemented, finite volume scheme developed by M. Ender [98] was used to calculate the tortuosity. This scheme uses the same approach with the same equations, and more details can be found in Ref. [98]. Similar to the FEM model ParCell3D, each voxel of the reconstruction data is used as a cell (or the data can be up-sampled before calculation) for the finite volume calculations. After assembling the system of equations, it is solved using the MINRES solver from MATLAB [178, 179].

This finite volume scheme can calculate the tortuosity of structures larger than 10^9 voxels on a high-performance workstation with 384 GB RAM. This is even more than ParCell3D is able to cope with. But although both implementations use the same approach, a difference between the results of up to 10 % (depending on the structure) was found for our reconstructed microstructures. This deviation is mainly caused by the main difference between the two methods: the way the different methods discretize the structure. As mentioned before, the FEM model discretizes the structure by using cubes (the voxels of the reconstruction) and the solution is represented by its values on the nodes (the 8 corners of the voxels). Similarly to that, in the finite volume scheme the voxels are used as cells for the calculations. Each cell constitutes one node, on which the values of the solution are given. These differences are illustrated in Fig. 4.15. In this example, the structure given in Fig. 4.13b is discretized with increasing resolution and the tortuosity is calculated with both models (finite element and finite volume). At which point the results are scaled by dividing them by the analytical result. As can be seen, in the case of the finite volume method each cube constitutes a node and the current can flow only horizontally and vertically through the faces of the cubes to the next cube ("staircase effect" due to the step-like pattern of pixel and voxel based structures). In contrast, with the finite element method each edge of the cubes is a node. Thus the current can also flow diagonally between the nodes. Consequently, at low resolutions a much better result is obtained for this extreme case of a diagonal pathway. As resolution increases the finite volume method results will tend toward the correct value. Please note

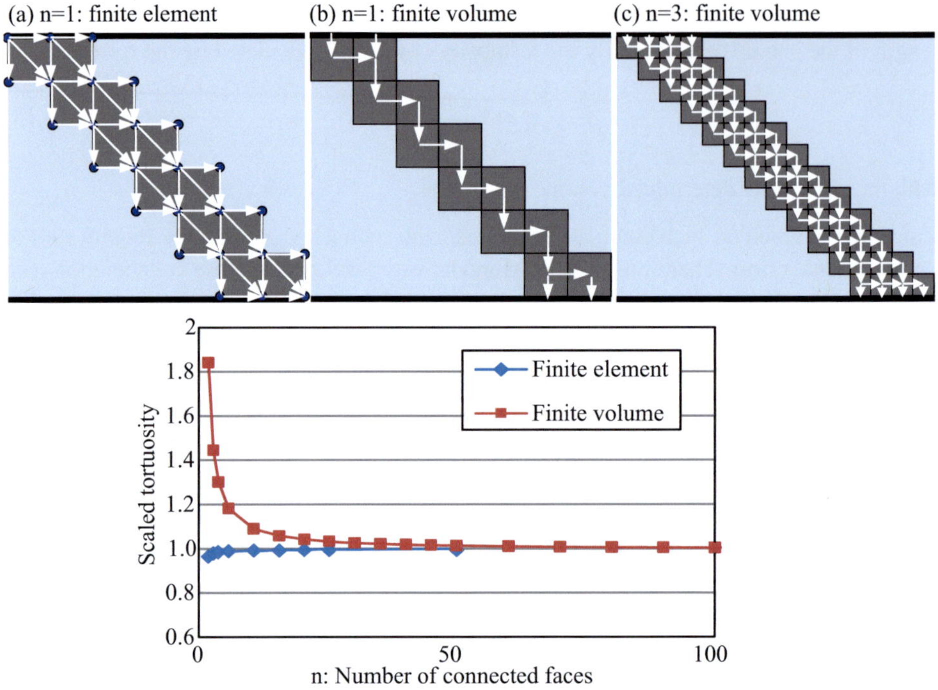

Figure 4.15.: Difference between the numerical models used to calculate the tortuosity: The example structure from Fig. 4.13b is discretized with increasing resolution. The arrows illustrate possible paths between the nodes. In this extreme case the finite volume method shows inferior results at low resolutions, as the current can only flow horizontally and vertically ("staircase effect"). In the case of finite elements a diagonal flow is also possible.

that the calculations were performed in 3D using cubes, although for clarity, Fig. 4.15 only provides 2D illustrations.

However, this example shows the case with the maximum difference between the two models. For structures with pathways closer to the vertical or horizontal the differences are much smaller. Moreover, looking at the differences between methods, a deviation of less than 10 % seems acceptable. It should also be mentioned that this method of numerically calculating the effective conductivity is considered to be the most reliable, because no *a priori* knowledge is needed. In contrast to tortuosity values obtained from geometrical approaches, the tortuosity obtained from calculating the effective conductivity can be used for the parametrization of homogenized models. As the finite volume model can cope with larger structures, and in order to enable a comparison of all tortuosity values within this thesis, all tortuosities were obtained using the finite volume scheme [98].

4.7.5. Particle and Pore Size Distribution

Another typical parameter for describing porous electrodes is the (average or median) particle size d_p and the particle size distribution of the materials present in the electrodes. Analogously, this holds for the pore size and pore size distribution, which is the reason why it is sometimes termed in literature as phase size distribution (PSD). For simplicity's sake, the term *particle size* will be used more often here, but this also includes the pores and the pore size distribution.

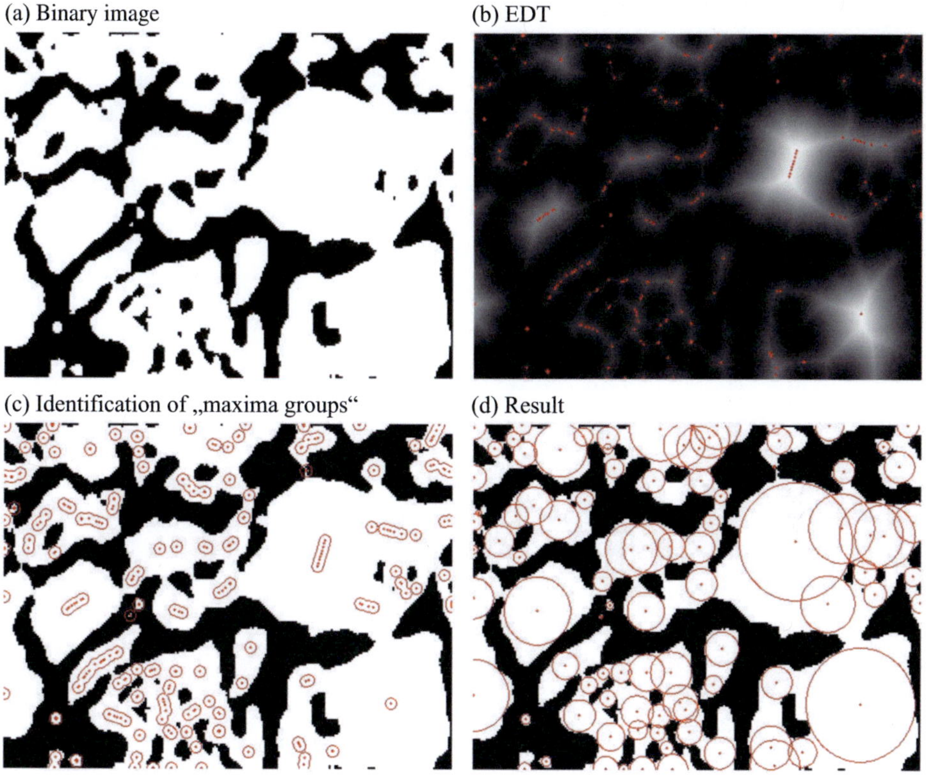

Figure 4.16.: 2D illustration of calculating the particle size distribution based on the Euclidean distance transformation (EDT): (a) Binary image and (b) distance map of a porous electrode. (c) Neighbouring maxima are merged if their distance is smaller than a given value. (d) The results are the largest spheres that can be placed within the particles (taken from [173]).

Although the particle size seems to be a simple parameter, its determination is far from trivial. This is mainly because the pores and particles inside a SOFC electrode structure are not equally sized and shaped, nor are they separated. For irregularly shaped and connected ("overlapping") particles it is hard to decide on the size of a particle, and it is not possible to describe such irregularly shaped particles correctly with parameters.

Nevertheless, it is possible to approximate these irregularly shaped particles in order to characterise and compare different structures or phases, knowing that an exact identification of the particle size and shape is impossible. Almost every group dealing with the quantification of SOFC electrodes has a different method of determining particle size. These methods can be very different and can therefore lead to very different results, even for the same or similar structures. This work uses an algorithm for the determination of the particle size distribution developed by M. Ender [98]. The calculation is based on the Euclidean distance transform (EDT), and its following description is mainly taken from Refs. [121, 173, 180].

The calculation is performed with binary images where all voxels of the investigated phase are labelled "1" while all other voxels are labelled "0". Applying the EDT to a 3D image (Fig. 4.16a) calculates for each value 1 voxel the Euclidean distance to the closest voxel with value 0. This results in a 3D distance map (Fig. 4.16b) where the value of each voxel corresponds to the distance to the closest voxel outside the particle. Hence, a local maximum in the distance map corresponds to the radius of the largest sphere than can be placed inside the particle. Since multiple maxima can occur within a particle it is necessary to detect these multiple maxima (Fig. 4.16c) and merge them to ensure a robust calculation of the particle size. This is done by disregarding all but the largest value if several maxima are within a defined distance. The sizes of the spheres defined by the remaining maxima (Fig. 4.16d) are taken as a measurement of the particle sizes, thus representing a lower bound for it. Thereby, two aspects are to be considered. First, the value of the maximum depends slightly on the position of the particle relative to the voxel grid. Second, the maxima overestimate the particle radius. Thus, a linear correction function is applied to calculate the particle sizes (diameters) d_p from the calculated maxima m_EDT, whereas the parameters of this function are determined by analysing a set of test structures consisting of non-overlapping spheres (cf. Refs. [98, 180])

$$d_\mathrm{p} = 2 \cdot (1.0132 \cdot m_\mathrm{EDT} + 0.3017). \tag{4.14}$$

The resulting set of individual particle sizes are used to calculate the particle size distribution of the investigated phase, as well as the average and median particle sizes.

A disadvantage of the method is that pores or cracks inside a particle will disturb the result, as the distance to the closest voxel of another face determines the maxima. Even one incorrectly assigned voxel inside a particle will result in the detection of multiple smaller particles around this voxel, instead of detecting one large particle. Therefore, it is of importance to correct segmentation errors using morphological filters (cf. Section 2.6) before calculating the particle sizes. Naturally, an insufficient resolution and an overly small reconstruction volume will lead to errors when calculating the PSD. These aspects will be discussed, together with all other parameters, in the next section.

4.8. Accuracy Aspects of Parameter Identification

This section presents and discusses studies on aspects related to the accuracy of parameter identification. It focuses on the main sources of error. This includes the resolution, the segmentation process and the reconstruction volume and their influence on the calculated parameters. The main purpose is to assess the calculated parameters from 3D reconstructions with respect to result accuracy. The secondary goal is to provide guidelines for accurate reconstruction and quantification of porous electrodes.

4.8.1. Resolution

The resolution of the 3D image data is one of the most important FIB tomography accuracy issues. Naturally, the resolution depends on the feature sizes of the structures. Hence the most favourable resolution factor is directly correlated to the particle size of the electrodes. In the following the resolution is expressed as the number of voxels per particle diameter.

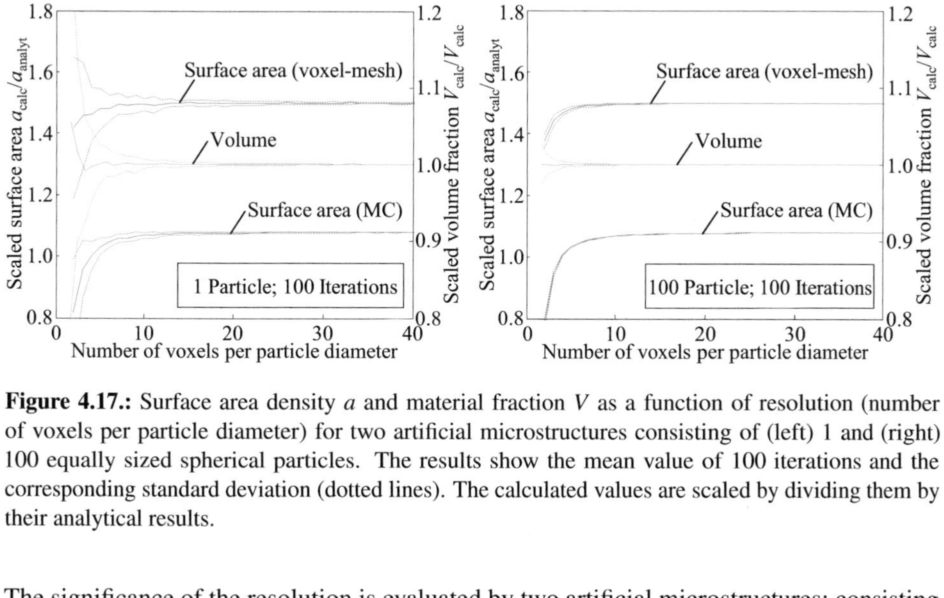

Figure 4.17.: Surface area density a and material fraction V as a function of resolution (number of voxels per particle diameter) for two artificial microstructures consisting of (left) 1 and (right) 100 equally sized spherical particles. The results show the mean value of 100 iterations and the corresponding standard deviation (dotted lines). The calculated values are scaled by dividing them by their analytical results.

The significance of the resolution is evaluated by two artificial microstructures; consisting of one and 100 equally sized and non-overlapping spherical particles in a set volume. The centres of the spherical particles were distributed randomly throughout the specified volume. Both volumes were fragmented into voxels, while an algorithm decided whether or not a voxel belongs to a particle (if more than 50 % of a voxel belongs to the particle). The surface area density a and the material fraction V of the particles were calculated (a_{calc} and V_{calc}, respectively) and scaled by dividing these values by their analytical ("correct") values

(a_{analyt} and V_{analyt}, respectively). It is therefore easy to quantify the error, e.g. a value of 1.2 would mean that the calculated parameter is overestimated by 20 %. The obtained results were plotted as a function of resolution (number of voxels per particle diameter) in Fig. 4.17. The surface area densities are given in grey (calculated directly from the voxel-mesh) and blue (calculated with the marching cubes algorithm), while the volume fraction is dispayed in green.

Since the position of the sphere-centres relative to the voxel-mesh influences the results, 100 arbitrary realizations of the microstructures are analysed for each structure. The results show the mean value over 100 iterations and their standard deviation (dashed lines). Naturally, for the higher number of particles (100 instead of one), the standard deviation decreases because the averaging over the particles already provides good statistics.

It can also be seen from Fig. 4.17 that the scaled volume quickly converges with the correct (normalized) value, 1, as resolution increases. This means that even a relatively low resolution can correctly reflect the volume fractions and that they are not very sensitive to resolution. In contrast, even at a high resolution the surface area of rounded particles will be overestimated, especially if approximated by voxels: by 50 % for the voxel based calculation and by 7 % if using the marching cubes algorithm (see Section 4.7.3). It can be seen that 10 to 15 voxels per particle-diameter is sufficient for converging results, depending on the number of particles inside the reconstructed volume. Below this threshold the resolution is too low and the surface area will be underestimated. Hence, the voxel size has to be accordingly small, with at least 10 voxels corresponding to the diameters of the particles. As a simple example, 10 voxels to a particle size of 500 nm means that the pixel size should be 50 nm or less. Furthermore, since the image dataset has to provide details in all three directions, the distance between the images (resolution in z-direction) should be matched to the lateral x–y image resolution. This minimal resolution of 10 voxels per particle diameter is in perfect agreement with the findings of Holzer *et al.* [181].

4.8.2. Re-sampling or Image Scaling

During image processing of the consecutive SEM images, resizing the 3D image data is a common process. One reason for this (which is referred to as *re-sampling*) is that in most reconstructions reported in literature the SEM image pixel size and the distance between the images are not equal (cf. Table 4.1). Most often it is the image distance (resolution in slicing direction) that shows lower resolution than the image resolution, aiming to save time by acquiring fewer images. Hence the data set is constituted of non-cubic voxels. But cubic voxels are sometimes advantageous for further data analysis (e.g. when calculating surface area with the marching cubes method). Therefore the pixel size has to be adjusted to the image distance, or vice versa. If the number of voxels is reduced by a re-sampling step, it is denoted as *down-sampling*, while a proliferation of voxels is denoted as *up-sampling*.

Table 4.6.: Microstructural parameters calculated from a sub-volume (size: 480 × 480 × 480 voxel) of cathode 2_{AP} (original data before re-sampling). These values were used to normalize the results in Figs. 4.18 and 4.19.

Original Data (size: 480 × 480 × 480 voxel)	LSCF	Pore
Material Fraction	44.7 %	55.3 %
Surface Area Density a	2.85 μm^{-1}	
Tortuosity τ	1.69	2.01
Average Particle/Pore Size d_p	557 nm	472 nm
Percolation “Sure”	99.99 %	99.51 %
Percolation “Unknown”	0.003 %	0.033 %
Percolation “Isolated”	0.003 %	0.455 %

Another common reason for re-sampling is that the data set can consist of too many voxels. This can either mean that not enough memory is available for further treatment or that some operations would need an inordinate amount of time. For example, many groups down-sample the data before they use it for simulation (e.g. [126, 128, 143]). But does re-sampling have an influence on the structure and especially on the calculated microstructural parameters?

This will be evaluated using a sub-volume (size: 480^3 voxels, $16.8 \times 16.8 \times 16.8\,\mu m^3$) of cathode 2_{AP}. The pixel size and image distance for this dataset are equally large, resulting in cubic voxels of $35 \times 35 \times 35\,nm^3$. The microstructural parameters calculated from the original data (before re-sampling) are listed in Table 4.6.

In the first study, the original data were down-sampled in several steps using the *imresize* function with the nearest-neighbour interpolation provided by MATLAB, and the effect on the calculated parameters was investigated. Details of the resultant data sets are listed in Table 4.7. The voxel size doubled with each down-sampling step, so that the number of voxels was reduced by a factor of 2^3 at each step. Note that the analysed volume remains the same and only the resolution (which is again expressed as the number of voxels per average particle diameter) gets coarser.

For each of the datasets the microstructural parameters pore fraction, surface area density, tortuosity of pores and LSCF, average particle- and pore size and the percolation shares (*sure*, *unknown* and *isolated*) were calculated. Similarly to the previous Section 4.8.1, the obtained results were normalized by dividing them by the results obtained from the original data (listed in Table 4.6) to clearly show how much the parameters change with the applied down-sampling. The normalized results are plotted in Fig. 4.18. The original data have a resolution of about 16 voxel per particle diameter, and with each down-sampling step the resolution is reduced by a factor of two. It can be seen that with further down-sampling (decreasing resolution) the difference increases relative to the values calculated from the original data in almost all cases. How strongly it does so, depends on the parameter, which becomes obvious when looking at the different scales of the y-axis.

Table 4.7.: Different steps of down-sampling the original data and resulting data sets with characteristic parameters. The resolution is expressed as the number of voxels per average particle diameter (which is 552 nm here).

Data Set	Volume / number of voxels	Voxel Size / nm^3	Volume / μm^3	Resolution / voxel per particle
Original Data	480^3	35^3	16.8^3	15.86
Re-sampling 1	240^3	70^3	16.8^3	7.93
Re-sampling 2	120^3	140^3	16.8^3	3.96
Re-sampling 3	60^3	280^3	16.8^3	1.98
Re-sampling 4	30^3	560^3	16.8^3	0.99
Re-sampling 5	15^3	1120^3	16.8^3	0.50
Re-sampling 6	8^3	2100^3	16.8^3	0.26

For example, the material fractions are almost unaffected by the down-sampling, which is unsurprising because the investigated structure has almost the same content of LSCF and pores. This ratio is not expected to change much with re-sampling. In contrast, the surface area density constantly decreases with decreasing resolution. This is mainly due to the fact that with decreasing resolution, small features or particles of the structure can no longer be accurately resolved. The detected surface area therefore decreases. Naturally, this influences the average phase size even more, which in these findings is the parameter with the highest sensitivity towards down-sampling here. Even after the first down-sampling the phase sizes had already increased by about 15 % compared to the original data. The pore- and material phase tortuosities increase with decreasing resolution. This can be explained by the fact that the tortuosity of a poorly resolved path increases, as discussed in Section 4.7.4 (cf. Fig. 4.15), and that small paths could no longer be resolved.

If looking at the percolation of LSCF and pores, it can be seen that the amount of percolating volume (labelled as "sure") barely changed, irrespective of the resolution. In contrast, the amounts of voxels which are labelled "unknown" or "isolated" seem to change dramatically with resolution. However, it should be noted that the plotted results are relative values (normalized by the calculated value of the original data) and that the amount of voxels with this status is negligible (between 0.003 and 0.455 %). Thus, even a small change in the absolute values leads to a large change in the relative values, although the absolute amount of unknown or isolated pores and particles is still very small.

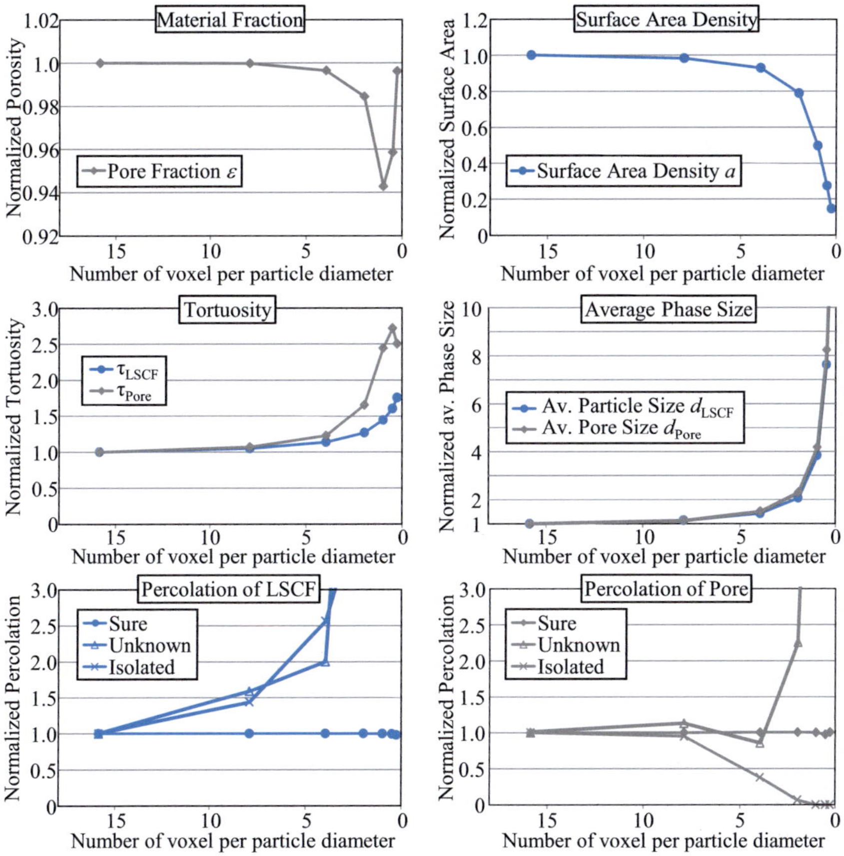

Figure 4.18.: Results of the down-sampling study. The calculated values for the different data sets are normalized by dividing them by the original data results (cf. Table 4.6).

Overall it must be concluded that a moderate down-sampling (by a factor of 2) only noticeably changed the average particle- and pore size (+15 %), and to a lesser extent the tortuosities of LSCF and pores (+4.8 % and +6.9 %). On the contrary, parameters like material fraction and surface area density were not affected very much (porosity -0.02 %, surface area density -1.5 %). A further decrease in resolution led to a strong change in the calculated parameters as the nature of the structure was changed. This should be avoided, even if the down-sampled data are used as a basis for simulation.

In the second study, the influence of non-cubic voxels on the parameters was investigated. The main idea here was to take every second (or later every third, fourth or fifth) image from the original data set, which resulted in non-cubic voxels of $35 \times 35 \times 70\,\mathrm{nm}^3$ size

($35 \times 35 \times 105\,nm^3$, $35 \times 35 \times 140\,nm^3$ or $35 \times 35 \times 175\,nm^3$, respectively). These data sets can then either be down-sampled (to volumes consisting of 240^3, 160^3, 120^3 or 96^3 voxels) or up-sampled (to volumes consisting of 480^3 voxels). Details of the resultant datasets are listed in Table 4.8. Afterwards the microstructural parameters could be calculated from these data sets and compared with the values calculated from the original data.

Table 4.8.: Different steps of up- and down-sampling and resulting data sets with characteristic parameters. The resolution is here expressed as the number of *images* (before re-sampling) per average particle diameter of the structure (which is 555 nm here).

Data Set	Volume / number of voxels	Voxel Size / nm^3	Resolution in z-direction / images per particle
Original Data	480^3	35^3	15.86
Up-sampling 1	480^3	35^3	7.93
Down-sampling 1	240^3	70^3	7.93
Up-sampling 2	480^3	35^3	5.29
Down-sampling 2	160^3	105^3	5.29
Up-sampling 3	480^3	35^3	3.96
Down-sampling 3	120^3	140^3	3.96
Up-sampling 4	480^3	35^3	3.17
Down-sampling 4	96^3	175^3	3.17

In accordance with the first study (cf. Fig. 4.18), the values calculated from the re-sampled data were normalized by dividing them by the original data results. The obtained results are now plotted as the number of *images* per average particle diameter. It should be noted, that for the case of up-sampling, only the data in z-direction (slicing direction) had to be up-sampled, as the SEM image resolution (x- and y-direction) was already sufficient with 480×480 pixels (35 nm pixel size) and hence a resolution of ~16 pixels per particle diameter in x- and y-direction. For the down-sampling, in contrast, only the resolution of the SEM images had to be adjusted (down-sampled) according to the image distance, resulting in the same resolution in all three directions.

The results are plotted in Fig. 4.19. It can be seen that the down-sampling results show the same trend for all parameters as in the first study of this section in Fig. 4.18. If comparing the absolute values it can be seen that the parameters change almost identically with decreasing resolution here, as they did for the down-sampling of the cubic voxel data in the first study. This is not surprising, since the same mechanisms discussed for the first study also hold true here. Nevertheless it shows that if the data are down-sampled before analysing them, there is almost no benefit from a higher resolution in slicing direction (larger number of images).

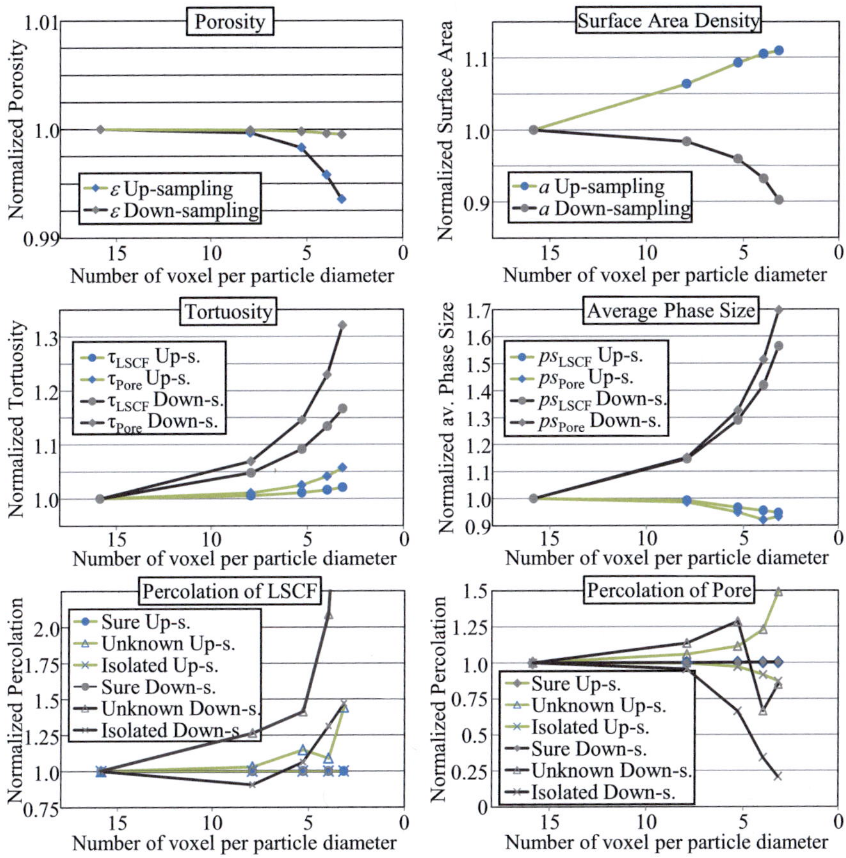

Figure 4.19.: Results of the up- and down-sampling study. The results were normalized by dividing them by the original data results.

For example, with a resolution of about 8 (or 4) voxels per particle diameter, the same values were calculated irrespective of whether every image is taken into account for down-sampling (first study) or only every second (or fourth) image. If the data are to be down-sampled anyway, it is not necessary to collect more images beforehand. Naturally, only acquiring half of the images would halve the data acquisition time. This is useful, especially for large reconstructed volumes, where the data acquisition time exceeds 24 hours.

As for the resolution in slicing direction, the same holds true for the resolution of the SEM images: there is no benefit to higher resolution SEM images, if the data are to be down-sampled afterwords. The error introduced by down-sampling can be very high, depending on the parameters and the degree of down-sampling.

But could some of the acquisition time be saved by acquiring fewer images (with a larger image distance), while still accurately calculating the parameters if the data were up-sampled? If looking at the results from the up-sampled data, the errors are much smaller than those from the down-sampled data. The already small error in calculating the porosity from the down-sampled data has almost disappeared. This is not surprising, as material fractions can be determined very precisely even from one large 2D image. The underestimation of the surface area density is only about half that of down-sampling. The tortuosities of LSCF and pore phase are much smaller if calculated from the up-sampled data than from the down-sampled ones. For a resolution of about eight images per particle diameter, the tortuosity is overestimated by only about 0.5 % for the LSCF phase and by 1.0 % for the pore phase (2.1 % for LSCF and 5.7 % for pores at a resolution of only ~3 images per particle). Also the average phase size changes only slightly with -0.8 % for the average particle size and -1.4 % for the average pore size at a resolution of 8 images per particle.

It can be concluded that the influence of re-sampling strongly depends on the parameter under investigation and the degree of re-sampling, which is clearly correlated to the resolution of the data (cf. Section 4.8.1). The material fractions have a very low sensitivity to re-sampling, while the other parameters clearly changed with re-sampling (and thus resolution). As discussed in Section 4.8.1, the minimum resolution should be 10 – 15 voxels per particle diameter. However, in the case of moderate up-sampling before parameter calculation (if the resolution is poorer in one direction), the resulting uncertainty in calculating microstructural parameters seems acceptable. Lower resolution is preferable in slicing direction, as collecting fewer images saves more time than lowering the resolution of the SEM images. Nevertheless, with decreasing resolution the error of the calculated parameters increases.

4.8.3. Threshold Value and Misalignment

As discussed in Section 4.6.1, identification of the true threshold value is of particular importance for a precise determination of the microstructural parameters. But how much do the parameters change if the right threshold value is not applied? This will be investigated in this section by looking at a ~840 μm^3 large volume of cathode type 1, together with the influence of the alignment procedure on its parameters. The exact influence of the alignment depends on the nature of the structure and on the image quality. Nevertheless, the following results also give a useful broad estimation of the sensitivities for other structures. Note, that the histogram of this dataset is displayed in Fig. 4.7 (red line, volume 1) and the true threshold value was identified as 97 (Section 4.6.1).

First, the influence of a varying threshold value used for greyscale segmentation is investigated on the parameters porosity fraction ε, surface area density a, tortuosity τ and average phase size d_p. All parameters are calculated for threshold values from 60 to 120 and their course is plotted in dependency to the threshold value in Fig. 4.20. As expected, the porosity is highly sensitive to the chosen threshold value, since with increasing threshold value more and more voxels are attributed to pores. Thus the porosity fraction clearly increases. With a threshold of 60 the porosity is about 31 %, whereas with a threshold of 120 it increases to about 55 %. It can be concluded that it is essential to establish the true threshold value for the calculation of material fractions.

In contrast, the surface area density shows very low sensitivity to a variation of the threshold value, as it only varies between 4.29 and 4.52 μm^{-1}. This is not surprising, as in this range a change in the threshold value means that the borders between pores and particles are slightly shifted (cf. Fig. 4.8).

The tortuosities of LSCF and pores also show clear dependency to the applied threshold value, which is naturally linked to the phase fractions. An increasing phase fraction means that existing paths enlarge and that more transport paths result, which leads to a lower tortuosity. Hence the tortuosity of the pore phase decreases (from 3.97 at a threshold value of 60 to 1.89 at a threshold value of 120), as the porosity increases. On the contrary, the LSCF tortuosity constantly increases with increasing threshold value (from 1.58 to 2.52), as the LSCF fraction is decreasing.

A similar behaviour can be seen for the average phase sizes, although the sensitivity is slightly lower than for the phase fractions and tortuosities. The pore size constantly increases, while the particle size is decreasing. Naturally, this is also linked to the increasing porosity and decreasing LSCF fraction with increasing threshold value.

Another important issue that can be grasped from Fig. 4.20 is the influence of the alignment procedure on the calculated parameters. As already discussed in Section 4.4, there are systematic shifts (e.g. due to the fixed position of the detector relative to the imaged sectional plane, which is shifting over time due to the FIB cutting) and random shifts (e.g. due to charging and disorders), which are responsible for consecutive images being slightly shifted against each other. These shifts can be easily seen by rotating the 3D image data by 90°, thus looking at the x-z plane of the 3D data (reconstructed side view, shown in Fig. 4.20 on the right hand side). If the images are not accurately aligned, the surface in this side view is very rough. The surface area would be thus overestimated (shown in Fig. 4.20). The accurate alignment was performed with ImageJ software, as already explained in Section 4.4. In the presented example, the poor alignment took only the systematic shifts into account and not the random ones, which lead to an overestimation of the surface area density of about 5 to 6 %. In contrast, the porosity and tortuosities were almost unaffected by the alignment procedure. Besides this, only the average particle- and pore sizes are also affected, in the case of poorly aligned images the average phase sizes would be underestimated by about 2 to 4.5 %.

It can be concluded that the chosen threshold has a huge influence on the phase fractions and, therefore, on the tortuosity and phase sizes. Only the surface area shows low sensitivity to the applied threshold in the investigated range of 60 to 120. But although the surface area density is less sensitive to a variation in the threshold value, it is very sensitive to the quality of the alignment procedure. Moreover, the phase sizes are also sensitive to the alignment quality, while the porosity and the tortuosity do not show an appreciable dependency to the alignment procedure.

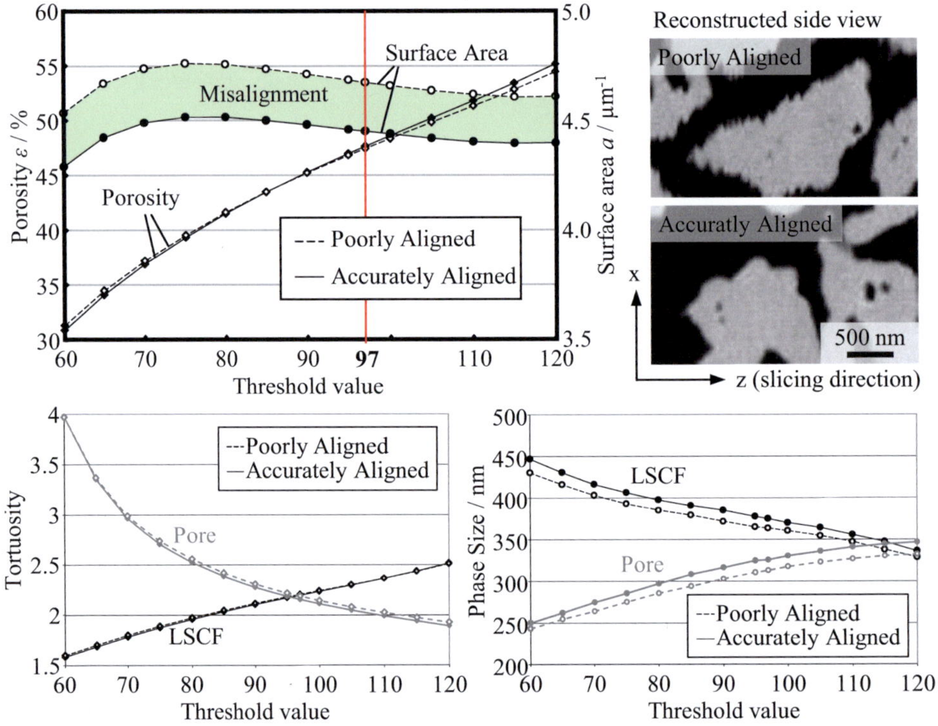

Figure 4.20.: Influence of threshold value and misalignment on the microstructural parameters for the example of cathode type 1: The surface area density is the only parameter which is clearly sensitive to misalignment (see also magnifications on the right hand side), while all other parameters (porosity, tortuosities and average phase sizes) are more sensitive towards the applied threshold value.

4.8.4. Size of Representative Volume Elements (RVE)

A very important aspect of microstructure reconstruction is the minimal volume size which has to be considered to be statistically representative for the whole structure. This section provides guidelines for defining such a representative volume element (RVE), as already introduced in Section 4.3. As mentioned there, the size of an RVE depends on different factors. First and foremost the feature sizes and the degree of structure homogenization have a huge influence. This alone makes it impossible to define a universal volume size as representative. Thus, such guidelines are only a suggestion for structures with a similar degree of homogenization, and they must be scaled with the particle size. Another important factor influencing the RVE size is the feature under investigation. Therefore, in the following the RVE size will be determined for different microstructural parameters.

Before defining the RVE sizes for a mixed conducting cathode, the meaning of such a representative volume should be briefly discussed. RVE means a sufficiently large volume, statistically representative of the whole structure [158], which also implies that the results do not depend on the exact location at which the volume is analyzed. This clearly shows the large impact of the degree of homogenization on the RVE size, and that it is almost impossible to define a RVE for inhomogeneous structures (e.g. graduated structures where the particle size is intentionally varied across the structure). Also, the RVE will never be exact unless there is some periodicity in the random structure, which is practically never the case for a real microstructure [157]. Moreover, the volume size obtained from FIB tomography is restricted: besides the challenge of operating a FIB/SEM system for several hours continuously (in our case normally for more than 24 h), a long run-time can lead to gradients of luminosity and contrast inside the 3D image data. Furthermore the resolution and hence the field of view in the SEM images is a limiting factor [180], so that one might have to work with volumes smaller than a RVE. Therefore, a more quantitative definition of the RVE is used here, similar to Kanit *et al.* [158], based on statistical arguments: within a RVE, the estimated properties (e.g. surface area, material fractions, tortuosity, area specific resistance, etc.) must ensure a given accuracy in a given volume. In case it is not possible to calculate a sufficiently large RVE, the accuracy must be assured by averaging over several smaller volumes of the microstructure to get the same accuracy. Therefore, please note that the average of several smaller volumes does not necessarily have to yield the same results as obtained by a sufficiently large RVE, as the influence of boundary effects increases. If a reconstructed volume is used for simulation, it is of particular importance to choose the right boundary conditions [85]. However, if the RVE is large enough, the response of the RVE must be independent from the type of boundary condition by definition [156, 158].

The minimum size of a RVE will be investigated in the following for a LSCF cathode. The parameters were first calculated from volumes, which were gradually increased (see illustration in Fig. 4.21a). Moreover, various volumes from different locations were analysed and the average value of these volumes was calculated, as well as the standard deviation (cf. Fig. 4.21b). The results are plotted as a function of the increasing volume in μm^3 (upper x-axes of the plots in Fig. 4.22). As the size of the RVE volume clearly depends on the feature sizes (e.g. average particle size of the structure), the lower x-axes shows the number of particles per volume edge length L_{VE}, which can be defined as

$$L_{VE} = \frac{\text{Volume Edge Length}\,[\mu m]}{\text{Particle Size } d_p\,[\mu m]} \tag{4.15}$$

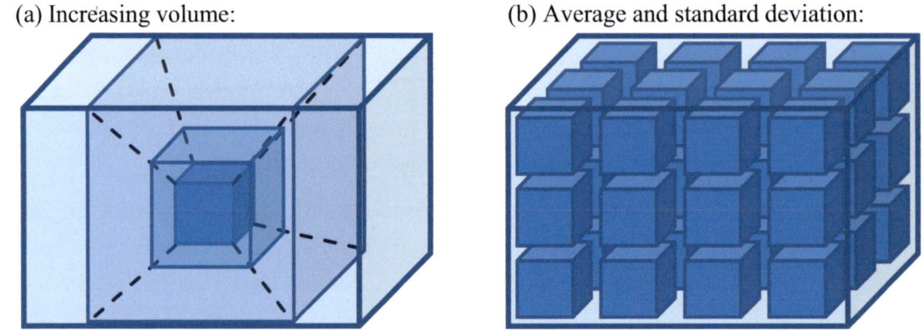

Figure 4.21.: Illustration of the study for determining the RVE size: (a) All parameters are calculated from a cubic volume which was gradually increased (solid line in Fig. 4.22). (b) Volumes at different locations are analysed and the average value (dashed line) together with its standard deviation (dotted line) is calculated and plotted in Fig. 4.22.

This dimensionless parameter provides an easily scalable measure of the size of an analysed cubic volume. This means that the volume at point L_{VE} contains $(L_{\mathrm{VE}})^3$ particles. Note, that all volumes here have a cubic geometry, in order to exclude any possible impact of the shape or aspect ratio of the volume on the parameters.

The size of a RVE for a two-phase cathode is determined by using cathode 2_{AP}, because this represents the standard cathode from FZJ and since this data set constitutes one of the largest volumes (total volume: $1250 \times 1150 \times 500$ voxels, $43.75 \times 40.25 \times 17.5\,\mu\mathrm{m}^3$). First, the porosity fraction, surface area density, tortuosity and average phase size are calculated from a cubic volume, which is incrementally increasing. The results (solid lines) are plotted in Fig. 4.22 as a function of the number of particles per volume edge length L_{VE} and hence the volume. As is clearly visible, all parameters change most at the beginning for small volumes and they converge at a certain volume to the final value. Thus, the volume where the values of the parameters do not change further with increasing volume can be defined as the RVE size.

The material fraction X_{LSCF} and surface area density a change most within a volume of fewer than 10^3 particles ($L_{\mathrm{VE}} < 10$), and less pronouncedly above. This holds especially true for the material fraction, which remained almost constantly above a volume of $L_{\mathrm{VE}} = 15$. Also the standard deviations of both parameters, calculated by analysing several equally-sized volumes (between 1000 volumes for $L_{\mathrm{VE}} = 3.2$ and 4 volumes for $L_{\mathrm{VE}} = 32$) at different locations, converge for volumes of $L_{\mathrm{VE}} = 10$ to 15. This means that a minimum of 10 to 15 particles per volume edge length (10^3 to 15^3 particles in total; hence a minimum volume of about 170 to $580\,\mu\mathrm{m}^3$) is necessary to obtain meaningful results for these parameters. However, the surface area density a still decreases softly until a volume of approximately $2500\,\mu\mathrm{m}^3$, which means that it is better to consider a much larger volume for calculating the surface area density.

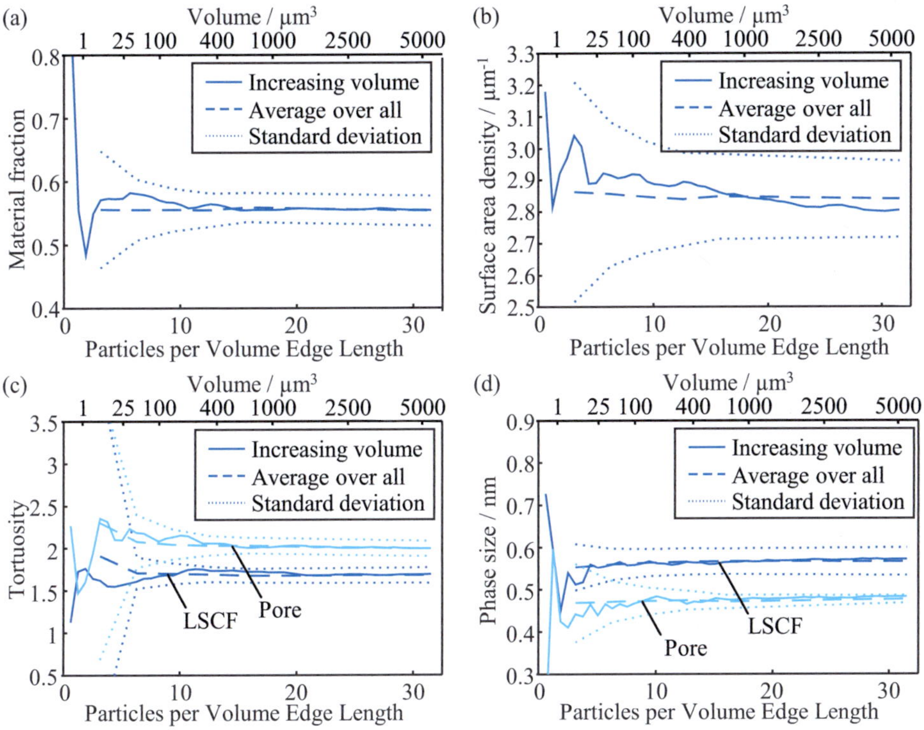

Figure 4.22.: Calculated microstructural parameters for an increasing cathode volume (solid lines): (a) material fraction, (b) surface area density, (c) tortuosities and (d) average phase sizes of the LSCF cathode type 2_{AP}. Moreover, the average value of several volumes is shown (dashed lines), together with the corresponding standard deviation (dotted lines).

Looking at the tortuosity values of LSCF and pores for the increasing volume, it can be seen that in the beginning the courses of both parameters go in opposite directions. This is clearly linked to the LSCF and pore fractions (compare Fig. 4.22a): A high LSCF material fraction obviously leads to low tortuosity values of LSCF, while the tortuosity of the pore phase is much higher, because the pore fraction is much lower. However, for volumes of 10 to 12 particles per volume edge length (about 170 to 300 μm^3), the values converge and then change only slightly afterwords. It can also be seen that the standard deviations are very large below $L_{\mathrm{VE}} = 6$, but almost constant above $L_{\mathrm{VE}} = 10$. Thus it can be concluded that a volume of only 10^3 to 12^3 particles (170 to 300 μm^3) already provides accurate values for the tortuosity of this cathode.

Another interesting aspect seen in Fig. 4.22c is the course of the average value of both tortuosities. In contrast to the material fraction, the average value of 1000 volumes which contain about 3.2^3 particles is higher than the average value of 40 volumes with 16^3 particles.

This is due to boundary effects during the tortuosity calculation, which are not present while calculating the material fraction by simply counting voxels. After a volume of about $L_{VE} = 6$, the average value stays quiet constant, which conversely means that from here on the boundary effect influence is almost negligible. These results support findings presented in [118]. If looking at the surface area density, a similar but less pronounced decrease of the average value can be observed below $L_{VE} \approx 13$. This is also due to a small boundary effect of the *marching cubes* method used to calculate the surface area.

Meanwhile, the average values of the phase sizes stay almost constant for all volume sizes. This is mainly because the algorithm which determines the phase sizes does not consider pores or particles connected to the boundaries of the analysed volumes. This avoids boundary effects. The average particle and pore sizes quickly converge to their final value. For the average pore size, a volume of about 10^3 particles seems to be sufficient, while the value for the average LSCF particle size already converges at about $L_{VE} = 5$.

It should be noted that the percolation of material and pores showed almost no dependency to the volume size, which is not surprising as even for the total volume the percolation of both phases is nearly 100 %. For smaller volumes, the probability that all particles and pores are percolating is higher. Consequently, the percolation was nearly 100 %, irrespective of the analysed volume size.

It can be concluded that for this cathode a volume containing 15^3 particles (about 580 μm^3) or more provides good statistics. On the other side, the results show that this (industrially fabricated) cathode has a very homogeneous microstructure, as the parameter values converge relatively quickly. However, for less-homogeneous structures, much larger volumes must be reconstructed to obtain a representative volume element.

4.9. Results of Cathode Reconstruction via FIB Tomography

This section deals with the results from the microstructural characterisation of the cathodes. The values obtained from the nine different cathode structures are presented in three subsections, which correspond to the three different cathode types. First, the results of type 1 are presented, showing the difference between three sub-volumes. Afterwards the values of type 2 are discussed, with the main goal of validating if and how the microstructure of LSCF cathodes changes during operation. In the third study the microstructural evolution of LSCF cathodes during manufacturing is investigated by analysing the four cathodes of type 3, which were sintered at different temperatures.

Type 1

As mentioned before, the total reconstructed volume of cathode type 1 (2585 μm) is divided into three sub-volumes of about 862 μm each, which - in accordance with the results presented in Section 4.8.4 - was found to be representative in Ref. [118]. However, Table 4.9 provides the main parameters calculated for all three sub-volumes, as well as the average value with the corresponding sample standard deviation, even though only three different volumes are analysed here.

Table 4.9.: Values for the microstructural parameters calculated from the three volumes of cathode type 1, together with the average value and the sample standard deviation (SD).

Parameter	Vol. 1	Vol. 2	Vol. 3	SD	Average
Porosity Fraction ε / %	47.6	50.0	47.4	1.42	48.4
Connected Pores / %	99.8	99.9	99.8	0.04	99.8
Connected LSCF / %	99.9	99.9	100	0.03	99.9
Total Surface Area a_{total} / μm^{-1}	4.45	4.42	4.39	0.03	4.42
Connected Surface Area a_{con} / μm^{-1}	4.42	4.39	4.36	0.03	4.39
Isolated Surface Area a_{iso} / μm^{-1}	0.03	0.02	0.03	0.01	0.03
Average Particle Size d_{LSCF} / nm	375	368	382	6.64	375
Median Particle Size d_{LSCF} / nm	376	369	376	4.14	373
Average Pore Size d_{pore} / nm	326	339	327	7.10	331
Median Pore Size d_{pore} / nm	305	314	305	5.04	308
Tortuosity of LSCF τ_{LSCF}	2.20	2.37	2.18	0.11	2.25
Tortuosity of Pore τ_{Pore}	2.15	2.06	2.15	0.05	2.12

For the porosity fraction the average value of the three sub-volumes is 48.4 %, with a sample standard deviation (SD) of 1.42 %. This is on the upper boundary of what had been predicted. The porosity of technically relevant LSCF cathodes is typically between 25 and 55 %. The optimal ratio between porosity and LSCF is a compromise, mainly influenced by the need for good gas diffusion, high effective ionic (and electronic) conductivity, mechanical stability and a large surface area density. The latter is theoretically largest at a ratio of 50 : 50 (cf. Ref. [26] and Fig. 5.25). However, good gas diffusion demands a higher porosity fraction, but then both, high effective conductivity and mechanical stability demand a higher material fraction.

The percolation of LSCF and the pores are nearly perfect, which is not surprising for a structure consisting of two phases with almost the same proportions. Thus, the sample standard deviation is low (0.04 and 0.03 %). Note that the amounts missing to 100 % are parts labelled as "isolated" or "unknown" (see Section 4.7.1).

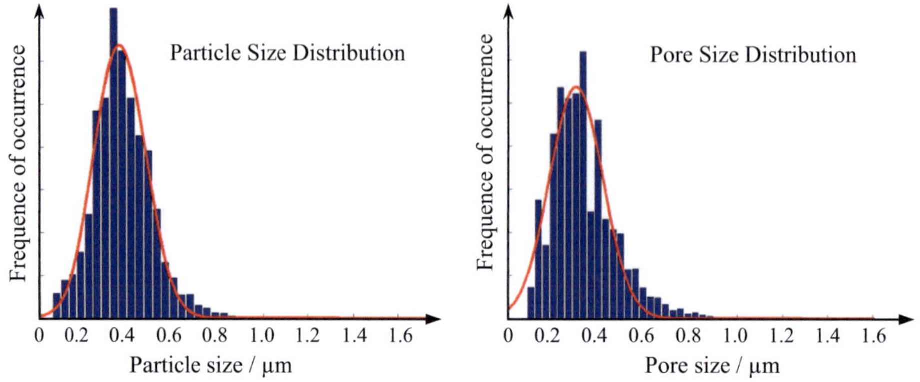

Figure 4.23.: Particle- and pore size distribution of volume 1 of cathode type 1. The (red) curved line shows an approximation of the distribution by a Gaussian function.

With an average value of 4.42 μm^{-1}, the surface area density a also shows a very low sample standard deviation (0.03 μm^{-1}), as the values of all three volumes are quite close to each other. This means that these parameters can be calculated from a volume of 862 µm with a very high accuracy. Moreover, only a very small part of the surface area is "isolated", while almost all is "connected". This is not surprising, knowing that almost all pores are connected.

The average and median phase sizes also show a relatively small sample standard deviation with less than 2 % difference from the average (and median) values. For the LSCF particle size, the average and median values are pretty close to each other with 375 and 373 nm. For the pore phase, the average value of 331 nm is slightly further off from the median value of 308 nm, but still close. If both characteristic values are close to each other, it is very likely that the particle size distribution is monomodal or even Gaussian shaped. The particle and pore size distributions calculated from volume 1 are shown in Fig. 4.23. Both phases have a monomodal distribution; especially the distributions of the LSCF particles show a Gaussian shape (indicated by the red line in Fig. 4.23). However, the phase size distributions of both phases are very similar.

For the tortuosity it can be observed that LSCF has a slightly higher average value of 2.25 compared to the tortuosity of the pores with 2.12. At first this is a little bit surprising, as the material fraction is larger than the pore fraction. The reason for this behaviour is that the LSCF network has more bottlenecks and dead ends compared to the pore phase. Despite this, the values for both phases are similar (about 6 % difference) and their sample standard deviations are also small with 0.11 for LSCF and 0.05 for the pores. Again, volume 2 shows a slightly different behaviour from the other two volumes. Volume 2 has a higher porosity fraction, especially in the middle of the volume, leading to a lower pore-tortuosity. This leads to bottlenecks for the LSCF phase and thus to a higher tortuosity. However, as already shown in Ref. [118], when accurately determining tortuosity it is important to consider a sufficient electrode height, which is obviously fulfilled here (see also [118]). Figure 4.24 shows the potential distribution inside the reconstructed volume 1 during the calculation of the tortuosities of both phases.

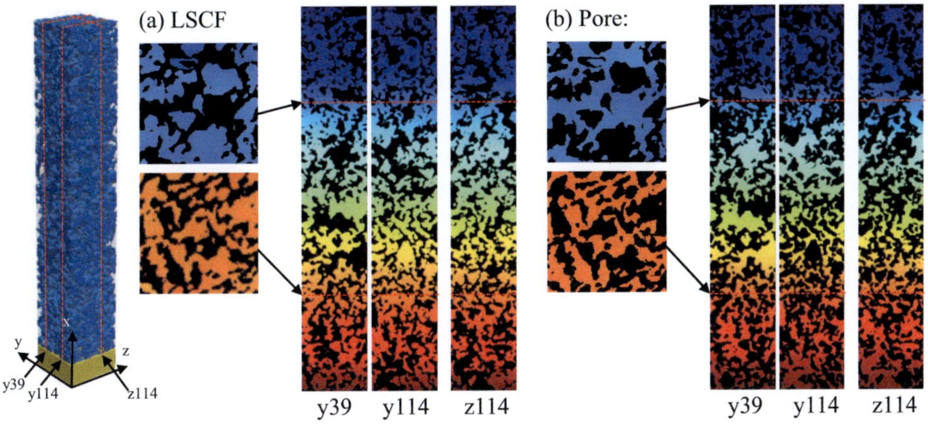

Figure 4.24.: Example of the potential distribution for volume 1 during the calculations of the tortuosities of LSCF and pores.

Altogether, it can be seen that the differences between the volumes are not negligible, but relatively low. The sample standard deviations are small, with less than 3 % from the average values for all parameters, except for τ_{LSCF} with 4.9 %. Considering the RVE definition used here ("the estimated properties must ensure a given accuracy in a given volume"), it can be concluded that the volumes are large enough to be RVEs. However, in further studies a larger volume would be useful to decrease the sample standard deviation of the parameters even more and thus determine them with even higher accuracy.

Type 2

Four identically fabricated type 2 cathodes were reconstructed and analysed. Three of them were operated for over 1000 h at 600°, 750° and 900 °C before reconstruction, while the fourth one was not operated (cf. Section 3.3). The idea was to quantify all four cathodes and compare the results of the operated cells with those from the "as prepared" (AP)-cell. It can thus be clarified for the first time if (and how) the microstructure of the LSCF cathode changes over time in relation to the operating conditions. In Section 5.4, the results will be used to identify if (and how) the material characteristics change over time.

Table 4.10.: Microstructural parameters calculated from the four cathodes of type 2.

Parameter	Type 2_{AP}	Type 2_{600}	Type 2_{750}	Type 2_{900}
Porosity Fraction ε / %	44.6	45.6	44.3	44.4
Connected Pores / %	99.56	99.69	99.66	99.61
Connected LSCF / %	100	100	99.99	99.99
Total Surface Area a_{total} / µm^{-1}	2.88	2.94	2.88	2.98
Connected Surface Area a_{con} / µm^{-1}	2.84	2.91	2.84	2.94
Isolated Surface Area a_{iso} / µm^{-1}	0.04	0.03	0.03	0.03
Average Particle Size d_{LSCF} / nm	552	539	556	543
Median Particle Size d_{LSCF} / nm	542	533	551	533
Average Pore Size d_{pore} / nm	449	450	452	438
Median Pore Size d_{pore} / nm	428	422	429	403
Tortuosity of LSCF τ_{LSCF}	1.68	1.71	1.68	1.73
Tortuosity of Pore τ_{Pore}	2.04	1.96	2.06	2.08

The results of the microstructural quantification are listed in Table 4.10. The porosity fractions of all four cells are in good agreement; only the porosity of cathode 2_{600} is about 1 % higher. This was expected, as the analysed volumes are very large (cf. Table 4.5), and the porosity and material fractions are not expected to change noticeably during operation. Moreover, the porosity is slightly lower than for cathode type 1, which might be due to the higher sintering temperature ($T_{\mathrm{sinter,type_2}} = 1080\,°\mathrm{C}$ compared to $T_{\mathrm{sinter,type_1}} = 1040\,°\mathrm{C}$). For all four structures, porosity is quite constant over the entire cathode thickness, as shown for cathode 2_{900} in Fig. 4.25. Furthermore, the LSCF and pore phase both show high connectivity, which also does not change markedly during operation.

The values also do not vary much for the total surface area density. From the lowest to the highest value, the difference is less than 3.5 %. This also holds for the amount of connected and isolated surface areas.

The scatter between the different cathode structures of both average and median phase sizes is slightly higher, but with about 6 % difference not very large. The LSCF particles are somewhat larger than the pores, but all values are in good agreement. It can be seen that the average and median values are relatively close to each other for both phases in all four cells. The particle and pore size distributions of all cells are almost identical, as can be seen in Fig. 4.26. The distributions of the particles are somewhat narrower than the pore size distributions, but between the different cathodes the differences are almost negligible, which clearly shows the high reproducibility of these structures.

As with all other parameters, the tortuosity values of LSCF and pores do not vary much between the different structures. A maximum difference of about 3 % for the LSCF and 6 % for the pores show that the tortuosities do not significantly change during operation.

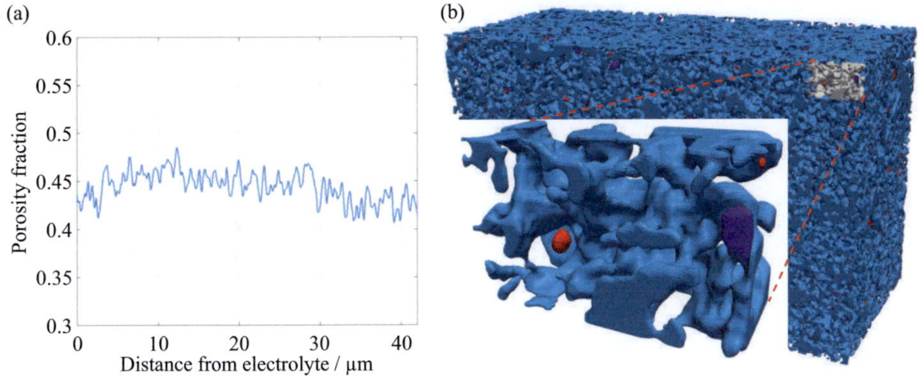

Figure 4.25.: (a) Distribution of porosity over the entire cathode thickness (from electrolyte to the current collector) for the example of cathode 2_{900}. (b) Results of the connectivity analysis of the pore phase, where red represents isolated parts, violet represents parts of unknown status and blue represents parts that are connected.

It can be concluded, that the structural differences between the four cathodes are very small. Moreover, no parameter shows temperature-linked decreases or increases. The scatter of the values for the different parameters lies within the manufacturing variation range. Degradation due to structural changes can be excluded.

Type 3

The goal of this study is to quantify the influence of sintering temperature T_{sinter} on the LSCF microstructure and predict the sintering behaviour in dependence to T_{sinter} (cf. Section 3.3). Four cathode structures of type 3, identically fabricated except for the sintering temperature, are reconstructed and analysed (cf. Fig. 4.27). A short introduction to the sintering process is given in Appendix C. Table 4.11 lists the results obtained from the four cathodes sintered at 960°, 1030°, 1080° and 1200 °C (each for 3 h). Note, that the standard sintering temperature at FZJ is 1080 °C.

With increasing sintering temperature, the porosity fraction of the four differently sintered cathodes clearly decreases. This trend is consistent with the physics of sintering (cf. Appendix C). However, the two structures sintered at the lowest temperatures show identical porosity fraction. This indicates that the expected densification is unpronounced below 1030 °C. Nevertheless, it is possible that cathode 3_{1030} has a lower overall porosity fraction than cathode 3_{960}, but due to small differences in manufacturing and/or due to small inhomogeneities of the reconstructed locations, the values were identical. However, due to the large volumes analysed, the differences should, in any case, be small.

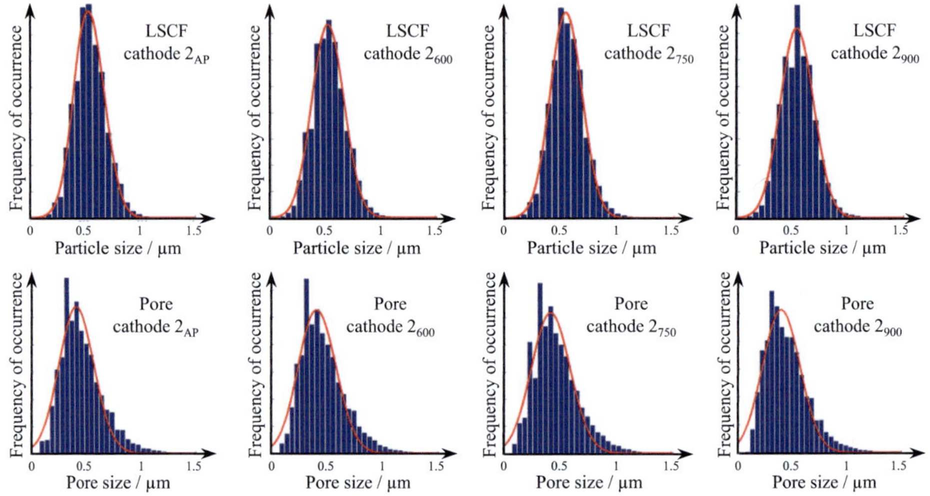

Figure 4.26.: Particle- and pore size distribution of all four cathodes of type 2. The (red) curved lines show approximations of the distributions by Gaussian functions.

As for all cathodes analysed within this thesis, the connectivity of both phases is almost ideal. This also holds for cathode 3_{1200} with only about 1.36 % isolated (and 1.97 % unknown) pores, even if the total porosity fraction is only 24.9 %. With this relatively low number of pores, more isolated pores were expected, as this amount is below the percolation threshold for a randomly distributed phase (see e.g. Ref. [26]).

The total surface area density decreases with increasing sintering temperature, which goes hand in hand with the fusion and increase of particles. Small particles give a larger surface area density than large particles. Additionally, the fusion of particles decreases the available surface area. The reduction of surface area density is quite large. From sintering temperatures of 960 °C to 1030 °C, the surface area density decreases by about 20 % from 6.10 μm^{-1} to 4.89 μm^{-1}. At the same time the average particle size increases by about the same amount from 253 nm to 304 nm. This clearly shows that the sintering process changes the structure markedly, even below 1030 °C. A further increase of the sintering temperature increases these trends for the surface area density and the phase sizes. At a temperature of 1080 °C, the surface area decreased to 3.28 μm^{-1}, while the average particle size increased to 452 nm. Increasing the sintering temperature to 1200 °C lead to a dramatic drop of the surface area density to only 0.81 μm^{-1}. This is again due to the extreme increase of the average particle size to 1553 nm, together with the decrease of porosity to only 24.9 %. The pore sizes increase similarly to the particle sizes. However, due to the strong increase in material density, the LSCF particle size increases much faster than the pore size.

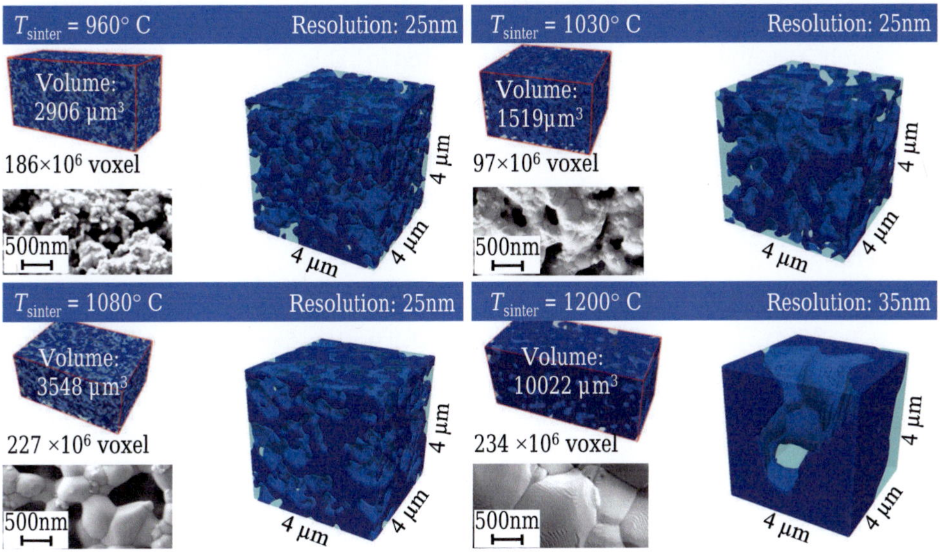

Figure 4.27.: The four cathode structures of type 3, sintered at different temperatures: for each cell, a part of a fractured cross-section SEM image is shown, together with the full reconstructed volume and a magnification of the reconstruction.

The LSCF tortuosity decreases from 2.88 to 1.23. Naturally, this is due to the increase of material density and the larger pathways, caused by the fused and increased particles. The opposite trend can be seen for the tortuosity of the pores. For sintering temperatures between 960° and 1080 °C, the tortuosity of the pores is almost constant between 2.00 and 2.04, while for 1200 °C it is almost twice as large with 3.94. It seems that up to 1080 °C, the simultaneously decrease of porosity and increasing pore sizes (better transport paths) results in an almost constant tortuosity in this temperature range. But with the extreme drop of porosity, many pathways are lost and thus the tortuosity increases.

It should be noted that Chen *et al.* [137] recently observed similar trends in a comparable study. For example, the tortuosity of LSCF decreased slightly from 1.4 to 1.1, while the porosity fraction simultaneously decreased from 47 % at 900 °C to only 15 % at 1200 °C. The tortuosity of pores remained almost constant between 900° and 1000 °C, but increased at 1100 °C and even more so at 1200 °C to a value of 2.5. The particle size (which they described as "the equivalent spherical diameter" and thus should be comparable to our method) is stated to have increased from 470 nm to 1700 nm, which is remarkably close to the results calculated in this work.

Overall, it can be concluded that the behaviour of the parameters with increasing sintering temperature meet expectations (considering the physics of sintering). The surface area density shows an especially large dependency to the sintering temperature. If one were to look purely at the microstructural parameters given in Table 4.11, it would seem that the lowest sintering temperature produces the best structures (as the surface area density is of special importance for the cathode performance). However, structural strength must also be considered, which naturally increases with increasing sintering temperature. Even if the determination of the structural strength is beyond the scope of this work, it should be mentioned that parts of cathode 3_{960} crumbled after the electrochemical characterisation (cf. Ref. [116]).

Table 4.11.: Microstructural parameters calculated from the four cathodes of type 3.

Parameter	Type 3_{960}	Type 3_{1030}	Type 3_{1080}	Type 3_{1200}
Porosity Fraction ε / %	53.3	53.3	46.7	24.9
Connected Pores / %	99.86	99.88	99.62	96.67
Connected LSCF / %	99.97	99.97	99.94	100
Total Surface Area a_{total} / µm^{-1}	6.10	4.89	3.28	0.81
Connected Surf. Area a_{con} / µm^{-1}	6.07	4.86	3.23	0.75
Isolated Surface Area a_{iso} / µm^{-1}	0.03	0.02	0.04	0.04
Average Particle Size d_{LSCF} / nm	253	304	452	1553
Median Particle Size d_{LSCF} / nm	242	288	451	1529
Average Pore Size d_{pore} / nm	247	307	376	815
Median Pore Size d_{pore} / nm	224	278	339	805
Tortuosity of LSCF τ_{LSCF}	2.88	2.43	2.04	1.23
Tortuosity of Pore τ_{pore}	2.01	2.00	2.04	3.94

5. Modelling and Simulation of Mixed Ionic/Electronic Conducting (MIEC) Cathodes

This chapter deals with the modelling and simulation of MIEC cathodes. Literature has presented myriads of performance models related to solid oxide fuel cells and their electrodes. These models can be classified according to their purposes and level of modelling detail [182]. Figure 5.1 shows one possible classification of SOFC related models [26], which span from models describing the system behaviour to models investigating the individual electrochemical processes in elementary kinetic reaction models. The level of modelling detail increases from system models up to reaction kinetic models, while the length-scale decreases. Depending on the modelling detail level, researchers may require different approaches, equations and parameters.

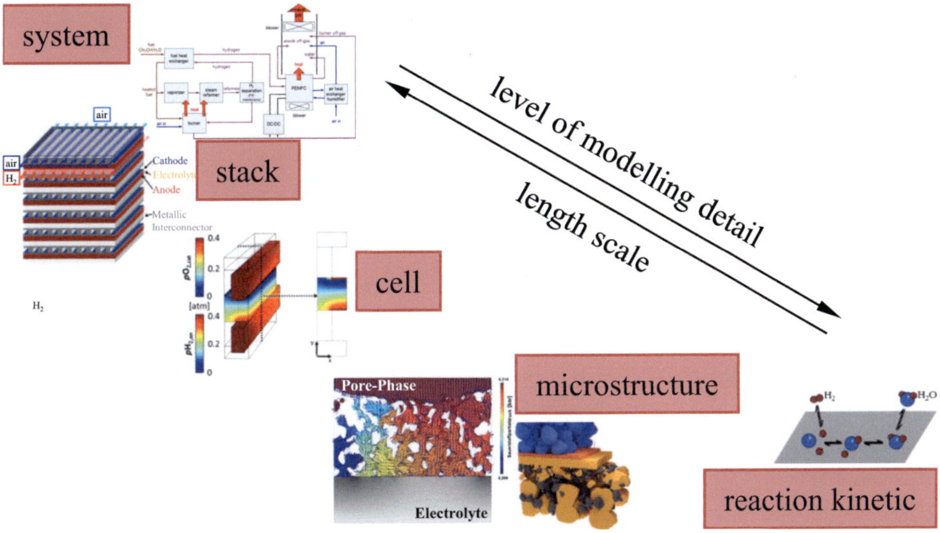

Figure 5.1.: Different levels of modelling fuel cells (adapted from Ref. [26]).

Since one of the goals of this work is to predict the performance of SOFC electrodes in dependency to microstructure (and the material) the focus of this paper now turns to the level of *microstructure modelling*. An overview of electrochemical performance models published in literature, with the main focus on models for MIEC cathodes, is given at the beginning of this chapter (Section 5.1). Three specific models for MIEC cathodes will be discussed in great detail: (1) the well-established Adler-Lane-Steel (ALS) model [60], in which all characteristics of the microstructure are represented in terms of volume-specific parameters (a homogenized or analytical model); (2) a FEM model from Rüger [26], which considers a simplified 3D representation of the electrode microstructure. It has to be mentioned that the work of Rüger can be seen, in several respects, as a forerunner to this thesis (in Section 5.2 a further development of the Rüger model is presented); (3) a model from Matsuzaki *et al.* [128] based on lattice Lattice Boltzmann Method (LBM) using reconstructed cathodes as computational domain.

After presenting the models, the first results of the high-resolution 3D FEM model are introduced in Section 5.3, where the influence of the material (represented by the material parameters k^{δ} and D^{δ} in the model) on the cathode performance will be studied. This will lead directly to the discussion of the need for an accurate and reliable method of determining these parameters, which will be shown in Section 5.4.

In Section 5.5, several accuracy aspects that arise during the calculation of the ASR_{cat} are discussed. This includes a detailed discussion on the minimal size of a RVE for the calculation of the area specific cathode resistance. Afterwards, the influence of $p\text{O}_2$ dependent k^{δ} and D^{δ} values (in contrast to constant values) inside the cathode microstructure is investigated. Finally, the Rüger model and ParCell3D are compared.

The influence of microstructure on electrode performance is investigated in Section 5.6. This will illustrate the huge impact of microstructure on cathode performance and thus the potential for the electrode performance optimization.

Stochastic microstructure models will be the topic of Section 5.7. These models generate artificial, synthetic microstructures with defined microstructural characteristics, which can then be used as computational domain in adequate 3D performance models. They can also be used to extract microstructural parameters. A model developed in this work is presented, capable of creating realistic synthetic microstructures. In Section 5.8, the stochastic microstructure model (in combination with ParCell3D and the ALS model) will be used to demonstrate how much cathode performance can be increased by optimizing the microstructure. Beneficial microstructural characteristics will be identified, thus providing guidelines for the structural optimization of MIEC cathodes.

5.1. State of the Art: Modelling the Performance of SOFC Electrodes

Many different electrochemical performance models for SOFC electrodes were presented over the last decades, an overview is given first. Furthermore, three models will be discussed in greater detail at the end of this section. The different approaches vary extremely as to their degree of complexity and the way they represent electrode microstructure. From a historical point of view, experimentally derived data were not available before X-ray or FIB tomography were established (cf. Section 4.1) so stochastic methods had to be used to create models capable of generating synthetic microstructures (cf. Section 5.7) and estimating microstructural parameters. Recently, increasing computational power and memory, with the availability of detailed realistic microstructures, has allowed performance models to become more complex and detailed.

5.1.1. Overview on Electrochemical Performance Models

Following Kreller *et al.* [183], performance models for SOFC electrodes can be divided into three "classes", depending on how they represent microstructure. The first class involves homogenized or analytical models (e.g. the ALS model [60]), where the microstructure is represented in terms of volume-specific macrohomogeneous parameters like porosity, volume-specific surface area and tortuosity. These models work well under many circumstances for homogeneous structures (provided that the required microstructural parameters are known, which are often difficult to measure). They do, however, become inadequate if the size of the active region is in the range (or smaller) of microstructural features (like the particle or pore size) [183]. Moreover, factors like local inhomogeneities, current constrictions, graduated structures or anisotropy in the structure cannot be considered by homogenized models. Models for dense thin film electrodes such as published by Jamnik and Maier [184], can be added to this class as well.

The second class contains models which include a simplified microstructure. This class includes (1) the FEM model published by Fleig and Maier [185] where the microstructure is considered as rods of cylindrically shaped particles; (2) the model of Lu and Adler [69, 186] which also uses cylindrical rods; and (3) the FEM model of Rüger [26], where the particles of the microstructure are represented by symmetrically aligned and equally sized cubes. Please note, that all the models mentioned thus far in this section are specifically for MIEC cathodes. However, most of the models presented in literature consider composite electrodes like LSM/YSZ cathodes or Ni/YSZ anodes. The second class also includes many models which are based on Monte Carlo packings of spherical particles, like the approaches presented by Schneider *et al.* [187, 188] where the model geometry is discretised into a resistance network and solved using Kirchhoff's law, or the approach presented by Cai *et al.* [189] employing the volume-of-fluid (VOF) method.

Although these models can capture some effects of 3D microstructures and work even when the size of the utilization region is competitive with the microstructure, they unavoidably leave many aspects of the microstructure untreated. The models often only consider homogeneous, monodispersed particles and neglect many details of the structure, like interfacial contact areas or contact angle [183]. The surface area density will normally be overestimated if using spheres and many of the models (like the rod-based models) ignore the tortuosity of the phases.

The third and last class of performance models use detailed, real microstructures as model geometry, obtained via, e.g. FIB or X-ray tomography. The potential to combine real microstructural data with numerical simulations opens many new perspectives. This approach has the advantage that all microstructural details (e.g. real shape and size of the particles, electrochemical hot spots, current constrictions, gradients inside the structure, anisotropy of the structure, etc.) are implicitly included in the model. Thus, experimentally derived data can be used to validate the electrochemical performance models. This has the potential to be applied as a powerful design tool for SOFC electrodes (cf. Section 5.8). An overview of published real microstructure based models is given in the following.

In 2008 Suzue *et al.* [190] performed the first simulation of a 3D microstructure, although not a real microstructure, but a stochastically reconstructed Ni/YSZ anode (a reconstruction based on a single 2D image). The LBM was applied to solve the species transport coupled with the electrochemical reactions. One year later Wilson *et al.* [123] used microstructural parameters obtained from a reconstruction of a LSM/YSZ cathode obtained from FIB tomography to predict the current-voltage behaviour.

However, in 2010 Shearing *et al.* [152] and Shikazono *et al.* [126] were among the first to perform electrochemical simulations based on a 3D reconstructed electrode (Ni/YSZ anodes, reconstructed by FIB tomography). Shearing used the individual voxels of the 3D image data to serve as finite volumes in the numerical discretization of a finite volume method (FVM) approach. In this model the active TPB length served as electrochemically active sites for current generation, following the Butler-Volmer equation. The anode overpotential was calculated as a function of the current density and temperature, and the resulting values were used to estimate the *ASR*. The length specific current density was used in the simulations as a fitting parameter to match experimental and simulated results. A good agreement was found between the applied current density per unit TPB and experimentally derived values from Ni pattern electrodes [78].

Shikazono *et al.* [126] published a similar model approach, but using LBM. The calculated overpotential for the Ni/YSZ anode agreed well with experimental data at a fuel supply of 1.2 % H_2O–98.8 % H_2, while the simulations slightly overestimated the experimental results at 10 % H_2O–90 % H_2.

One year later, Matsuzaki *et al.* [128] from the same group presented a model for MIEC cathodes, which was also based on the LBM and expressed the reaction current by a Butler–Volmer like equation. As this is one of the few studies on MIEC cathodes, this model will be discussed in greater detail at the end of this section.

Kishimoto *et al.* [191] used a model (based on FVM) similar to the ones presented by Shearing and Shikazono to simulate the anode polarization of a Ni/YSZ anode. Like almost all groups, they down-sampled the reconstruction data before simulation, but presented a new approach ("a sub-grid scale model") to keep the quality of the structural information, which is normally lost by "standard" down-sampling. It was stated that this approach could either improve the simulation accuracy under a given calculation grid system or reduce computational load for the same degree of simulation accuracy [191].

Based on a reconstruction of a porous single-phase LSM cathode, in 2013 Lynch *et al.* [192] presented a computational framework for the simulation of multiphysics phenomena on actual porous structures. This used the commercial software package COMSOL. The paper focusses more on the methodology than on the actual results, nonetheless, e.g. the current density of the cathode was calculated for varying cathodic overpotentials.

However, most of the models published in literature deal with Ni/YSZ anodes, while only few models consider MIEC cathodes. In the following, three models for the simulation of MIEC cathodes will be discussed in greater detail (one from each class discussed above).

5.1.2. Homogenized Analytical Model of Adler/Lane/Steel

The Adler-Lane-Steel (ALS) model [60] is a homogenized model which describes the impedance spectrum of a porous MIEC cathode. It considers a symmetrical cell with two identical cathodes (see Section 3.2), where the microstructure is represented by three microstructural parameters: porosity fraction ε, surface area density a and solid-phase tortuosity τ_{LSCF}. This model is also a good example of how the parameters calculated in Chapter 4 can be used in SOFC electrode modelling. The model consists of analytical expressions derived by solving a system of coupled partial differential equations, which describe e.g. the displacement of the oxygen vacancy concentration in the mixed conductor from its equilibrium value. The total impedance Z_{cat} of the symmetrical cell is assumed to be the sum of the individual impedances:

$$Z_{\text{cat}} = R_{\text{electrolyte}} + Z_{\text{CT}} + Z_{\text{chem}} \tag{5.1}$$

where $R_{\text{electrolyte}}$ is the (ohmic) electrolyte resistance, Z_{CT} is the impedance of the electron- and ion-transfer processes at the current collector/electrode and electrode/electrolyte interfaces and Z_{chem} is the convoluted impedance of the surface exchange and solid-state diffusion in the cathode material. Technically it would also include the gas diffusion losses in the pores of the cathode and the current collector, but this is often neglected for the sake of simplicity, as in the following explanation. The applied assumption of non-charge-transfer allows for a linearisation of the model equations, as well as other significant simplifications. Thus, generally, the coupled system of differential equations can be transferred to an analytical equation for the description of the chemical impedance

$$Z_{\text{chem}} = R_{\text{chem}} \sqrt{\frac{1}{1 + i\omega t_{\text{chem}}}} \cdot \left[\tanh\left(\frac{l_{\text{cat}}}{l_\delta} \sqrt{1 + iwt_{\text{chem}}} \right) \right]^{-1} \tag{5.2}$$

where l_{cat} is the cathode thickness, l_δ a characteristic length (called penetration depth) which will be described below, and R_{chem} and t_{chem} are the characteristic resistance and time constant of the chemical processes, respectively. They are related to the thermodynamic, surface kinetic, and transport properties of MIEC materials [60]:

$$R_{\text{chem}} = \left(\frac{RT}{2F^2} \right) \sqrt{\frac{\tau_{\text{LSCF}}}{(1-\varepsilon) c_{\text{v}} D_{\text{v}} a r_0 (\alpha_{\text{f}} + \alpha_{\text{b}})}} \tag{5.3}$$

$$t_{\text{chem}} = \frac{c_{\text{v}}(1-\varepsilon)}{\gamma_{\text{v}} a r_0 (\alpha_{\text{f}} + \alpha_{\text{b}})}. \tag{5.4}$$

Here, c_v is the vacancy concentration, γ_{v} is the thermodynamic factor for vacancies, D_{v} is the vacancy diffusion coefficient, r_0 is the exchange neutral flux density and α_{f} and α_{b} are kinetic parameters. Using $r_0 \cdot (\alpha_{\text{f}} + \alpha_{\text{b}}) = k^* \cdot c_{\text{mc}}$, these parameters can be converted into commonly used thermodynamical parameters. The thermodynamic factor γ_v basically describes the ability of the material to change the oxygen ion concentration depending on the external partial pressure. It is therefore closely linked to the chemical capacity ($C_{\text{chem}} = t_{\text{chem}}/R_{\text{chem}}$) [3]. It should be noted that the transport of oxygen ions in the ALS model is described in terms of oxygen vacancies as the mobile species (using D_{v}, k and γ_{v}). However, in the Rüger model [26] and the model presented in Section 5.2, the *chemical* diffusion coefficient D^δ and the associated surface exchange coefficient k^δ are used, together with the thermodynamic factor γ_{o} (see Appendix B for more information and the transformation between both sets of parameters).

The characteristic length l_δ (also called *penetration depth*) describes the height of the electrochemically active area and can be defined by the ionic current density. It corresponds to the cathode height at which the ion current has dropped to $1/e \approx 36.8\,\%$. Its value can be calculated by the following formula, which depends on material and microstructural characteristics

$$l_\delta = \sqrt{\frac{c_v D_v (1-\varepsilon)}{a r_0 (\alpha_f + \alpha_b) \tau_{\text{LSCF}}}} \approx \sqrt{\frac{(1-\varepsilon)}{a \cdot \tau_{\text{LSCF}}} \cdot \frac{D^\delta}{k^\delta}}. \tag{5.5}$$

It is worth noting that a prerequisite for the validity of the model is that the particle size of the considered cathode is much smaller than l_δ [60]. However, depending on the ratio between l_δ and the cathode thickness l_{cat}, two special cases can arise for Eq. (5.2), due to the course of the *tangens hyperbolicus* (tanh), as illustrated in Fig. 5.2.

In the case of a thick, "semi-infinite" cathode ($l_{cat} > 3 \cdot l_\delta$), $\tanh(l_{cat}/l_\delta)$ approaches unity and thus Eq. (5.2) reduces to

$$Z_{chem} = R_{chem}\sqrt{\frac{1}{1+i\omega t_{chem}}}. \tag{5.6}$$

In this case, where cathode thickness is much larger compared to the active area, the resistance (or rather the impedance) is determined by both the surface exchange and the solid-state diffusion ("mixed-controlled area"). Here, Eq. (5.6) corresponds to the Gerischer-impedance [193].

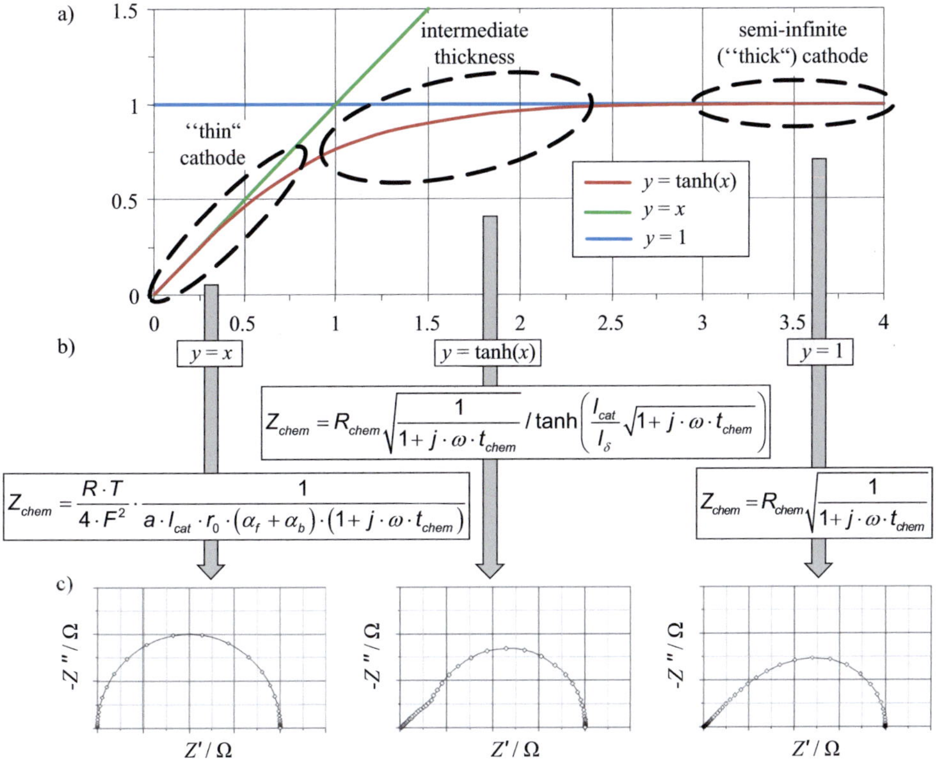

Figure 5.2.: (a) Plot of the functions $y = \tanh(x)$ and $y = x$ for $x = 0\ldots4$. (b) Case differentiation: Eq. (5.2) can be simplified by approximating tanh in dependency to l_{cat}/l_δ. (c) Nyquist-plots of the impedance's resulting from the three cases shown in (b) (adapted from [3]).

For the case of thinner cathodes ($l_{\text{cat}} < 3 \cdot l_\delta$), the shape of the impedance deviates more and more from the shape of the Gerischer-impedance, as illustrated in Fig. 5.2c. The characteristic slope of 45° for high frequencies caused by diffusion decreases with decreasing l_{cat}, until for $l_{\text{cat}} < l_\delta$ the impedance becomes a perfect semicircle (impedance of a RC element [3]). This is because the *tanh* term in Eq. (5.2) becomes linear, yielding an impedance which is independent of the diffusion

$$Z_{\text{chem}} = \frac{RT}{2F^2} \cdot \frac{1}{a l_{\text{cat}} r_0 (\alpha_f + \alpha_b)(1 + i\omega t_{\text{chem}})}. \tag{5.7}$$

Thus, the polarization resistance depends only on the surface exchange, which scales with the available surface area ($1 \,/\, a \cdot l_{\text{cat}}$). In other words, diffusion is negligible for these thin electrodes and the entire surface area of the electrode is utilized for O_2 reduction.

For cathodes with an intermediate thickness ($3 \cdot l_\delta > l_{\text{cat}} > l_\delta$), the impedance shows a quasi-semicircular impedance response with a slight diffusive part at high frequencies. The thinner the cathode, the weaker the diffusive behaviour is pronounced.

Please note, that this model can also be used to calculate the material parameters k^δ and D^δ by using results (R_{chem} and t_{chem}) obtained from EIS measurements, as will be shown in Section 5.4.

5.1.3. Rüger Model: Using Simplified Microstructures

The 3D FEM model developed by Rüger [26, 194] calculates the area specific resistance ASR_{cat} of MIEC cathodes as a function of material parameters and microstrcuture. The following processes are taken into account at a constant temperature T and a constant pressure p: in the cathode (i) gas diffusion, (ii) surface exchange, (iii) bulk diffusion and (iv) charge transfer, and in the electrolyte (v) the transport of the oxygen ions. However, the area specific resistance of the electrolyte ($ASR_{\text{elyt}} = l_{\text{elyt}}/\sigma_{\text{elyt}}$) is subtracted from the total ASR calculated by the model; it is therefore not included in the ASR_{cat}. Nevertheless, in contrast to homogenized or 1D models, current constrictions at the cathode/electrolyte interface are taken into account by the model and these losses are attributed to the ASR_{cat}. The equations used to model the processes are listed in Fig. 5.3. Because the model presented in the next section is an extension of this model and uses the same equations, the modelling approach will be discussed in greater detail in Sections 5.2, together with all relevant equations.

In contrast to the ALS model, the processes considered in this model are spatially resolved in a 3D microstructure. The actual microstructure of the electrode is taken into account by approximating the structural particles by cubes, which are symmetrically aligned and equally sized (see Fig. 5.4). It was shown in [26] that a base area (cubes in y- and z-direction) of at least 7×7 cubes delivers reliable results. The model is implemented in the finite element

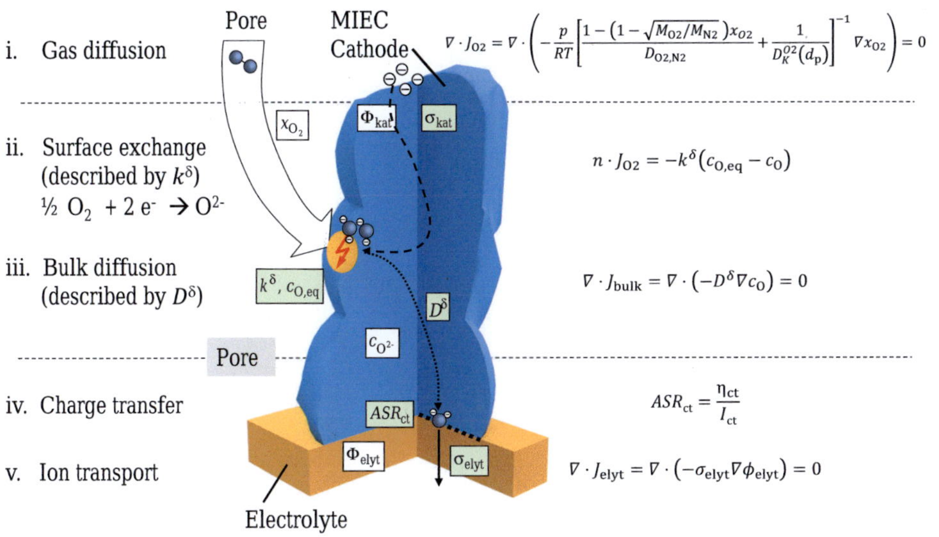

Figure 5.3.: Processes considered in the Rüger model: (i) gas diffusion, (ii) surface exchange, (iii) bulk diffusion, (iv) charge transfer and (v) ion transport in the electrolyte.

commercial software COMSOL Multiphysics (Version 3.4) [195]. For this version, the maximum number of cubes which can be considered for the simulations is about 2500.

As mentioned above, each cube (or voxel) of the model geometry represents one particle of the actual microstructure. Therefore the edge length of the cubes l_v corresponds to the mean particle size (particle diameter) d_{LSCF} of the actual particles. In addition, the mean particle size corresponds to the mean pore size ($d_{LSCF} = d_{pore}$), since all cubes do have the same size. The model geometry is automatically generated according to parameters such as cathode thickness l_{cat} and mean particle size. The symmetrically aligned cubes are randomly assigned to be (i) electrolyte material, (ii) MIEC material or (iii) pores, according to the desired material composition and overall porosity. The model is not restricted to pure MIEC cathodes, it can also consider composite cathodes like MIEC/CGO. A current collector is connected to the top of the cathode structure to collect and distribute current and/or gas to the porous cathode structure. An electrolyte is located underneath to collect current. As the (randomly) assigned material distribution can have a slight influence on the results (only relatively small volumes can be considered due to the limiting number of cubes), the average value of 15 to 60 simulations is typically used.

The material is considered according to three MIEC material parameters: the chemical diffusion coefficient D^δ, the surface exchange coefficient k^δ (cf. ALS model above), and the oxygen ion equilibrium concentration in the perovskite lattice $c_{O^{2-},eq}$. As with the ALS model, the material parameters D^{δ} and k^{δ} are taken from literature (for each temperature and oxygen partial pressure). Also the oxygen ion equilibrium concentration $c_{O^{2-},eq}$ was evaluated as a function of oxygen partial pressure for each temperature of measured data [196] as shown in [26]. It should be noted that the model does not contain adjustable parameters.

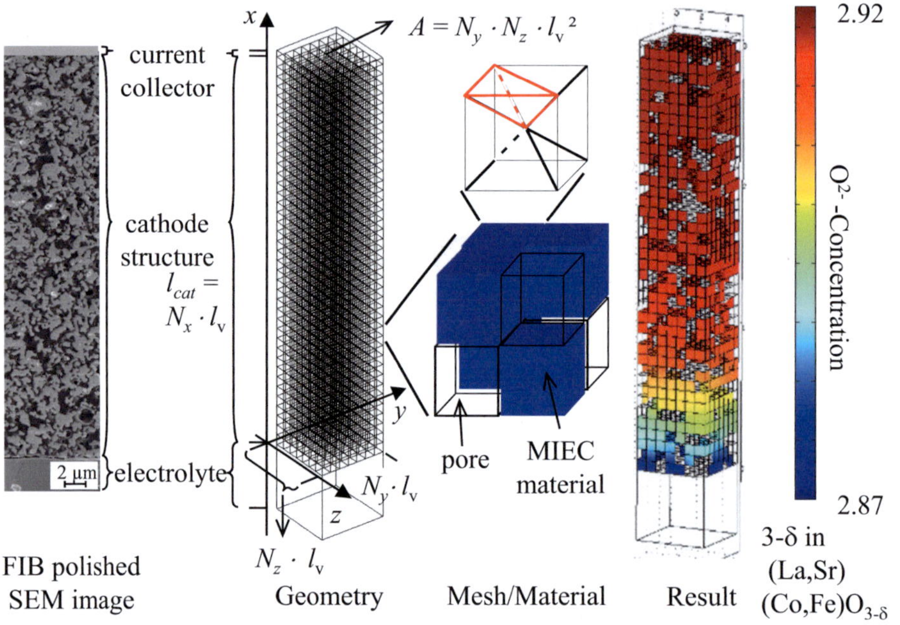

Figure 5.4.: 3D FEM model from Rüger: the actual microstructure is approximated by symmetrically aligned, equally sized cubes (middle). The model is solved by a commercial FEM software, an example result shows the spatial resolved O^{2-}-concentration in the material on the right hand side (adopted from [101]).

The greatest limitation of the model is the coarse approximation of more or less spherical cathode particles (grains) in the actual structure: because the COMSOL Multiphysics software has a limited number of cubes, it is not possible to approximate one particle by more than one cube *and* consider a representative number of particles. Furthermore, the model current pathways are enabled through the edges and corners of the cubes (e.g. for the material distribution in the centre of Fig. 5.4, a current pathway through the joint edge of the upper left and the lower right cubes is given). Only by a radical increase in the cube-number could the "true" size and shape of the particles be more accurately reproduced. To overcome the limitations of COMSOL and consider a significantly larger number of cubes (or voxels), the software *ParCell3D* was developed. With this software it is possible to directly use reconstructed electrodes as model geometry, as will be explained in Section 5.2.

It is worth mentioning that this model was extended in a master thesis supervised in the context of this work to allow for simulations in the time-domain. This enables the simulation of the impedance spectrum of a MIEC cathode [197].

5.1.4. Matsuzaki Model: Using Reconstructed Microstructures

The paper presented by Matsuzaki *et al.* [128] is one of the few publications that deals with MIEC cathode models, and that can consider a real microstructure for simulations. A LSCF cathode was reconstructed using FIB tomography and the obtained 3D data down-sampled to obtain cubic voxels for simulation. The implemented processes were solved in the 3D microstructure to calculate the cathode's overpotential by the Lattice Boltzmann Method, assuming local equilibrium in the solid oxide. The reaction current i_{reac} is expressed by a Buttler-Volmer-like equation. Surface electron transfer is assumed to be the rate limiting step [128]. The modelling approach will be discussed in detail below.

The diffusion of oxygen in the pores is modelled by the following equation, which is based on the dusty-gas model (DGM) [198], neglecting the convection and assuming a constant total pressure

$$\left(\left[\frac{1-(1-\sqrt{M_{\text{O}_2}/M_{\text{N}_2}})x_{\text{O}_2}}{D_{\text{O}_2,\text{N}_2}}+\frac{1}{D_{\text{O}_2,\text{k}}}\right]^{-1}\nabla c_{\text{O}_2}\right)=-\frac{1}{4F}i_{\text{reac}} \tag{5.8}$$

where x_{O_2} is the molar fraction of the oxygen, F the Faraday constant and M_i the molar mass of the gas component i. Then $D_{\text{O}_2,\text{N}_2}$ and $D_{\text{O}_2,\text{k}}$ are the binary and Knudsen diffusion coefficients, respectively, given as

$$D_{\text{O}_2,\text{N}_2}=0.018833\sqrt{\frac{1}{M_{\text{O}_2}}+\frac{1}{M_{\text{N}_2}}}\frac{T^{3/2}}{p\Omega_{\text{D}}\zeta^2_{\text{O}_2,\text{N}_2}} \tag{5.9}$$

$$D_{\text{O}_2,\text{k}}=\frac{1}{3}\sqrt{\frac{8RT}{\pi M_{\text{O}_2}}}d_{\text{pore}} \tag{5.10}$$

where Ω_{D} is the collision integral, $\zeta=(\zeta_{\text{O}}+\zeta_{\text{N}})/2$ is the intermolecular force constant, p the total pressure, R is the ideal gas constant and d_{pore} is the mean pore diameter (cf. Ref. [128]). It should be mentioned that the pores seem to be much narrower, with a mean diameter d_{pore} of only 178 nm (compared to cathode type 2_{AP} with 449 nm). The different values are not due to different methods used to determine d_{pore}, since their procedure is similar to the one presented in Section 4.7.

The transport of electrons and ions in the MIEC phase is modelled by the equations

$$\nabla\left(\frac{\sigma_{e^-}}{F}\nabla\tilde{\mu}_{e^-}\right) = -i_{reac}, \tag{5.11}$$

$$\nabla\left(\frac{\sigma_{O^{2-}}}{2F}\nabla\tilde{\mu}_{O^{2-}}\right) = i_{reac}, \tag{5.12}$$

where $\tilde{\mu}_{e^-}$, $\tilde{\mu}_{O^{2-}}$, σ_{e^-} and $\sigma_{O^{2-}}$ are electrochemical potentials and conductivities of electron and oxide ion, respectively. The electron conductivity σ_{e^-} and the chemical diffusion coefficient D^{δ} are extracted by fitting the experimental data (depending on T and p_{O_2}) as published by Bouwmeester *et al.* [199].

The electrochemical reaction is assumed to take place on the surface of the MIEC material (gas/solid interface), and electron transfer to adsorbed O_{ad} atom (O_{ad}+$e^- \rightarrow O^-_{ad}$) is assumed to be the rate limiting step for the MIEC cathode reaction. The reaction current at the gas/solid interface is modeled as a function of the activation overpotential η_{act} in the following Butler-Volmer-like equation, as e.g. proposed by Fleig [200]

$$i_{reac,2PB} = i_0 A_{2PB}\left\{\exp\left(\frac{1.2F}{RT}\eta_{act}\right) - \exp\left(-\frac{1.0F}{RT}\eta_{act}\right)\right\} \tag{5.13}$$

where the oxygen partial pressure dependency and the activation energy of the linear exchange current i_0 are assumed as

$$i_0 = 1.47\times 10^6 \cdot p_{O_2}^{0.2}\exp\left(-\frac{85859}{RT}\right). \tag{5.14}$$

Equation (5.14) is obtained from fitting experimental data published by Esquirol *et al.* [201]. The local activation overpotential η_{act} on the surface is defined as

$$\eta_{act} = -\frac{1}{2F}\left(2\tilde{\mu}_{e^-,MIEC} - \tilde{\mu}_{O^{2-},MIEC} + \frac{1}{2}RT\log p_{O_2,gas}\right). \tag{5.15}$$

Equations (5.8), (5.11) and (5.12) are solved using LBM. An electrolyte and a current collector layer are added to the top and bottom of the structure, respectively. Adiabatic boundary conditions are applied to the outer boundaries of the volume, while a constant gas composition (Dirichlet boundary) is given at the current collector surface. Constant electronic and ionic current flux conditions (Neumann boundary) are imposed on the current collector and electrolyte boundaries, respectively. On the solid surface in the porous media a zero-flux boundary is imposed [128]. The total cathode overpotential $\eta_{cathode}$ is obtained by subtracting the ohmic losses of current collector, electrolyte and reference electrode from the total overpotential, defined as the difference between EMF and terminal voltage [128].

In Ref. [128] simulations were performed in a volume of approximately 843 μm^3 (~4.5×10^6 voxel) for $P = 1.013 \times 10^5$ Pa at varying temperatures (973.15 K to 1073.15 K), current densities i (0.01 A/cm^2 to 0.2 A/cm^2) and gas compositions ($O_2 : N_2$; 100:0, 50:50 and 20:80). The simulated cathode overpotential were found to agree well with experimentally observed data, especially at high p_{O_2}. However, at O_2 = 20 %, T = 973 K, i = 0.2 A/cm^2, the cathode overpotential was significantly overestimated, which was attributed to the decline in oxygen ionic conductivity at low p_{O_2}.

5.2. High Resolution 3D FEM Model for Real Microstructures: ParCell3D

A FEM performance model is presented in this section, capable of calculating the area specific resistance of the cathode ASR_{cat} in dependence to real microstructures, i.e. obtained from 3D reconstructions. It is an extension of the Rüger model [26], where computational limits required a simplified microstructure (cf. Section 5.1). For this extension a new, in-house software called *ParCell3D* was developed in collaboration with T. Carraro (University of Heidelberg). ParCell3D efficiently solves the coupled system of equations defined by the model in a complex modelling domain by high performance computing (HPC) techniques, enabling it to solve models with up to 10^8 voxels. ParCell3D was written by T. Carraro as an objective oriented C++ code to make its extensions more flexible. The model was published in peer-reviewed journals [104, 119, 177, 194]. This section is based on those papers.

As a computational domain, the model includes the porous cathode (cat; e.g. obtained by FIB tomography), where a current collector/gas channel (CC) layer is connected on the top and an electrolyte (elyt) layer is present on the bottom (see Fig. 5.5). The anode is not present, but assumed as an ideal and lossless counter electrode (CE). The computational domain does not necessarily have to be a reconstructed real cathode, as shown in Fig. 5.6 where the oxygen ion concentration inside the MIEC phase of cathode type 1 is illustrated. In principle all geometries based on cubic voxels (or cubes, respectively) can be used.

The considered volume is divided into subregions corresponding to different materials: Ω_{pore} is the corresponding region of the pore phase, Ω_{MIEC} is the region of the MIEC bulk phase, while Ω_{elyt} is the electrolyte. Between these subregions the interfaces are defined: Γ_{PM} is the surface between MIEC phase and pore phase (thus equivalent to Γ_{MP}), Γ_{ME} (or Γ_{EM}) is the interface between MIEC phase and electrolyte and Γ_{PE} (or Γ_{EP}) is the interface between electrolyte and pores.

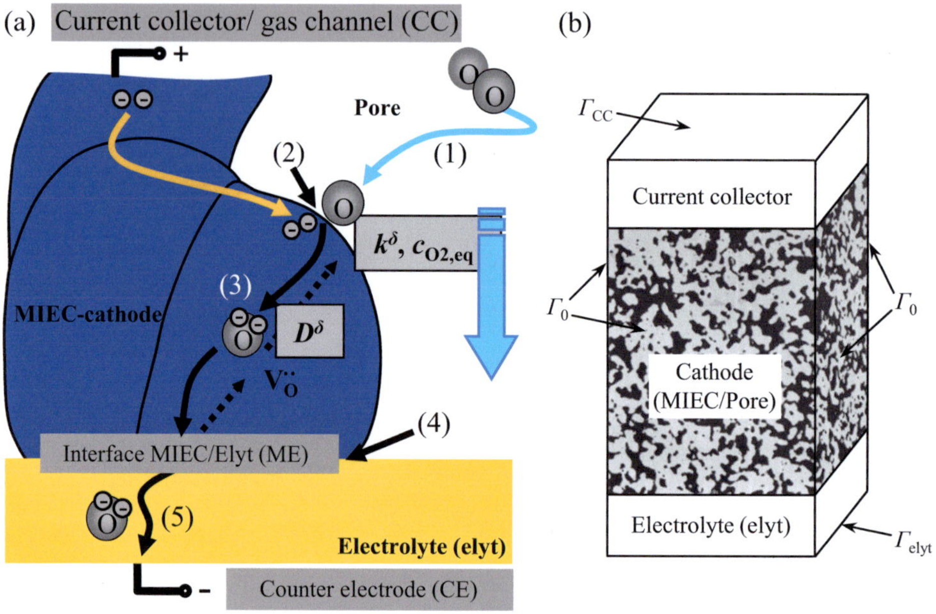

Figure 5.5.: (a) Scheme of model approach and processes considered in ParCell3D. (b) Schematic definition of the boundary surfaces for the application of boundary conditions.

5.2.1. Material Parameters

As mentioned in Section 5.1.3, the mixed conducting cathode material is described by three parameters: (a) the surface exchange coefficient k^δ, (b) the chemical diffusion coefficient D^δ and (c) the oxygen ion equilibrium concentration $c_{\mathrm{O}^{2-},\mathrm{eq}}$. The latter is taken from Ref. [26], where measurement data from Refs. [196, 199, 202] were fitted in the $p\mathrm{O}_2$ range between 10^{-4} and 1 bar, and the functional dependency on T and $p\mathrm{O}_2$ (measured in bar) of the type

$$c_{\mathrm{O}^{2-},\mathrm{eq}}(T, p_{\mathrm{O}_2}) = C_2(T)\log(p_{\mathrm{O}_2}) + C_1(T) \tag{5.16}$$

is given. The values for C_1 and C_2 from Ref. [26] are listed in Table 5.1 for $p_{\mathrm{O}_2} = 0.21$ bar. This table also contains two example literature value sets for k^δ and D^δ [2, 199]. It should be mentioned that the values from Ref. [2] are obtained by interpolating values published between $T = 571\,°\mathrm{C}$ and $870\,°\mathrm{C}$, whereas the values for D^δ had to be divided by a factor of four before interpolation, due to an incorrect assumption in Ref. [2] (see Eq. (6.17) and the explanation above the equation in Appendix B). For the simulations in this chapter, different values for k^δ and D^δ are used, which is also mentioned at the relevant passages. However, the scatter of k^δ and D^δ values published for the same materials is very large, as will be

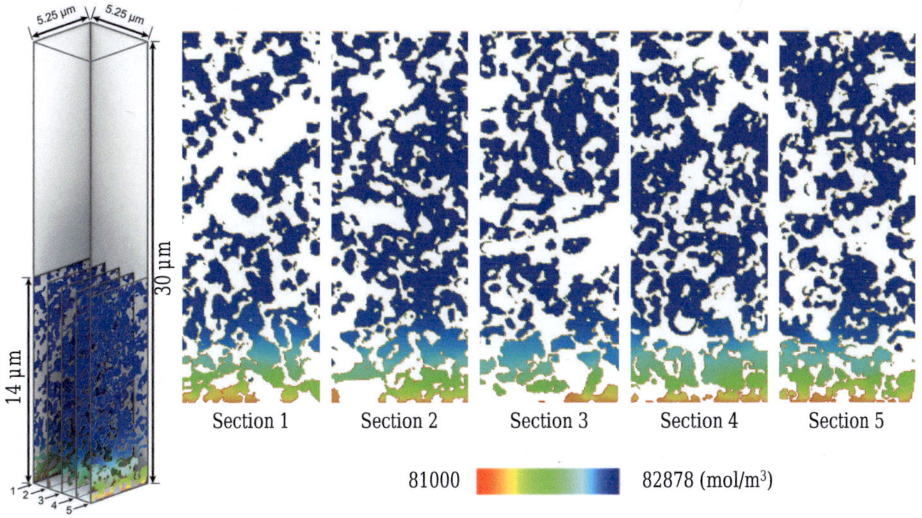

Figure 5.6.: Oxygen ion concentration (in mol/m^3) in the MIEC material for the example of cathode type 1. Calculated for $T = 750\,°\mathrm{C}$ using the material parameters from [2].

discussed in Section 5.3. In Section 5.4 a method is presented for the precise determination of these parameters by combining the results from FIB tomography, EIS measurements and microstructural modelling. The values from Ref. [2] shown in Table 5.1 will be examined and modified.

5.2.2. Working Principle

The working principle of the model is based on the actual physics: as boundary conditions the oxygen partial pressures on the anode (the counter electrode CE, respectively) and cathode side ($p_{O_2,\mathrm{cat}}$ and $p_{O_2,\mathrm{CE}}$) as well as the potentials of the electrodes ($\Phi_{\mathrm{MIEC}}, \Phi_{\mathrm{CE}}$) are applied to Γ_{CC} and Γ_{elyt}, respectively (cf. Fig. 5.5). Since the anode is assumed to be ideal and lossless, it follows that $\Phi_{\mathrm{CE}} = \Phi_{\mathrm{elyt}}$ and $p_{O_2,\mathrm{CE}} = p_{O_2,\mathrm{elyt}}$. The potential of the CE can be chosen arbitrarily and was set to 0 V. For calculating the cathode potential given by

$$\Phi_{\mathrm{MIEC}} = U_{\mathrm{N,cat}}(p_{O_2,\mathrm{cat}}, p_{O_2,\mathrm{elyt}}) + \Phi_{\mathrm{CE}} - \eta_{\mathrm{Model}}, \tag{5.17}$$

the total voltage losses (or overpotential) of the model η_{Model} occurring in the model are needed, which have to be defined. At $\eta_{\mathrm{Model}} = 0\,\mathrm{V}$, the current I is given as 0 A, whereas for $\eta_{\mathrm{Model}} > 0\,\mathrm{V}$, the current I is created and calculated by an integration of the current density at Γ_{elyt} (see Fig. 5.5). Thus, η_{Model} is the driving force of the model.

Table 5.1.: Material parameters at $p_{O_2} = 0.21\,\mathrm{bar}$ used for the simulations with ParCell3D. The coefficients C_1 and C_2 (as published in Ref. [26]) are used to calculate the oxygen ion concentration at equilibrium according to Eq. (5.16). Moreover, two example value sets for k^δ and D^δ are listed from Bouwmeester [199] and Leonide [2].

Temperature	$c_{O^{2-},eq}$		Bouwmeester		Leonide	
T / °C	C_1	C_2	k^δ / m s^{-1}	D^δ / m^{2}s^{-1}	k^δ / m s^{-1}	D^δ / m^{2}s^{-1}
600	365	85263	-	-	$2.30 \cdot 10^{-6}$	$2.71 \cdot 10^{-11}$
650	550	85083	$2.28 \cdot 10^{-6}$	$7.18 \cdot 10^{-11}$	$4.90 \cdot 10^{-6}$	$6.75 \cdot 10^{-11}$
700	700	84836	$7.04 \cdot 10^{-6}$	$2.04 \cdot 10^{-10}$	$9.64 \cdot 10^{-6}$	$1.53 \cdot 10^{-10}$
750	851	84410	$1.20 \cdot 10^{-5}$	$4.40 \cdot 10^{-10}$	$1.77 \cdot 10^{-5}$	$3.19 \cdot 10^{-10}$
800	976	83888	$1.50 \cdot 10^{-5}$	$7.32 \cdot 10^{-10}$	$3.09 \cdot 10^{-5}$	$6.22 \cdot 10^{-10}$

The area specific resistance of the model is calculated as $ASR_{\mathrm{Model}} = \eta_{\mathrm{Model}}/I$. Subtracting the specific resistance of the electrolyte ($ASR_{\mathrm{elyt}} = l_{\mathrm{elyt}}/\sigma_{\mathrm{elyt}}$), the area specific resistance of the cathode is given by $ASR_{\mathrm{cat}} = ASR_{\mathrm{Model}} - ASR_{\mathrm{elyt}}$.

As mentioned in Section 5.1, the model takes several processes into account at a constant temperature T and a constant pressure p [194]. These processes are (1) gas diffusion of oxygen in the porous structure (including a Knudsen term), (2) surface exchange (oxygen reduction) at the interface MIEC/pores, (3) bulk diffusion of oxygen ions in the lattice, (4) charge transfer at the interface cathode/electrolyte, and (5) ionic conduction in the electrolyte. These processes are described in the following.

5.2.3. Equations, Parameters and Boundary Conditions

Gas Diffusion

The gas in the pore phase is considered as a binary mixture of oxygen and nitrogen. The diffusion of oxygen molecules within the pores is modelled using the Dusty-Gas Model (DGM)

$$J_{O_2} = -\frac{p}{RT}\left[\frac{1-(1-\sqrt{M_{O_2}/M_{N_2}})x_{O_2}}{D_{O_2N_2}} + \frac{1}{D^k_{O_2}(d_{\mathrm{pore}})}\right]^{-1} \nabla x_{O_2} \quad \text{in } \Omega_{\mathrm{pore}}, \tag{5.18}$$

where x_{O_2} is the oxygen molar fraction in the pore phase; M_{O_2} and M_{N_2} are the atomic masses of the oxygen and nitrogen molecules, respectively; $D_{O_2N_2}$ is the binary diffusion coefficient and $D^k_{O_2}(d_{\mathrm{pore}})$ is the Knudsen diffusion coefficient, as in Eq. (5.10). The DGM was found to be the most appropriate model to simulate gas transport phenomena inside a SOFC electrode (in Ref. [198]) and it is based on the *Stefan-Maxwell equation*, in combination with the *Knudsen Diffusion*. This approach takes into account the interaction of oxygen molecules with the pore walls by $D^k_{O_2}(d_{pore})$, which is dependent on d_{pore}. The binary diffusion coefficient $D_{O_2N_2}$ is calculated after Fuller [203]. The gas diffusion equation is thus given as

$$\nabla \cdot J_{O_2} = 0 \quad \text{in } \Omega_{pore} \tag{5.19}$$

with boundary condition $x_{O_2} = \bar{x}_{O_2}$ on Γ_{CC} (see Fig. 5.5), which imposes the oxygen partial pressure at the interface with the gas channel, and $n \cdot J_{O_2} = 0$ on Γ_{PE} which represents an insulation condition between pores and electrolyte. An additional boundary condition between pores and MIEC is described in the next subsection.

Surface Exchange

Oxygen from the pore phase is adsorbed into the cathode material on the surface. By accepting electrons from the cathode, oxygen ions with a double negative charge are formed (O^{2-}). The speed of the surface exchange depends on both the temperature and the oxygen partial pressure and is described by the surface exchange coefficient k^δ. Without a current the oxygen ion concentration in the MIEC phase reaches an oxygen partial pressure dependent equilibrium value $c_{O^{2-},eq}$ [196]. This concentration decreases under current load ($c_{O^{2-}} < c_{O^{2-},eq}$) and an oxygen flow through the boundary takes place, which can be modelled by

$$n \cdot J_{O_2} = k^\delta \left(c_{O^{2-},eq} - c_{O^{2-}} \right) / 2 \quad \text{on } \Gamma_{PM} \tag{5.20a}$$

$$n \cdot J_{bulk} = -k^\delta \left(c_{O^{2-},eq} - c_{O^{2-}} \right) \quad \text{on } \Gamma_{MP}. \tag{5.20b}$$

Values for $c_{O^{2-},eq}$ are taken from experimental data published in Ref. [196]. In the Rüger model [26] and in the early version of ParCell3D, a constant value for k^δ (and also for D^δ) is used for the simulations, since the temperature is assumed to be constant and as the oxygen partial pressure in the pores hardly changes. However, in the current version of ParCell3D, k^δ (and D^δ) is pO_2 dependent, considered by

$$k^\delta = k^\delta_{ref} \cdot (pO_2 / pO_{2,ref})^{\alpha_k}, \tag{5.21}$$

where reference values for k^δ and pO_2 are used, and α_k is the slope of k^δ with varying pO_2. Nevertheless, this barely changes the results, as will be shown in Section 5.5.2. The boundary condition 5.20a is coupled with Eq. 5.19, while 5.20b is a boundary condition for the equation of bulk diffusion, which is described in the next subsection.

Bulk Diffusion

Due to high cathodic electronic conductivity, a constant electrical potential is assumed (e.g. at 800 °C the conductivity of the porous LSCF cathode is 8899 $\mathrm{S\,m^{-1}}$, which results in an ohmic loss of below 0.035 $\mathrm{m\Omega cm^2}$ for a cathode of 30 µm thickness, whereas the polarization losses exceed 10 $\mathrm{m\Omega cm^2}$). Its value is determined by the conditions on oxygen partial pressures and potentials on both electrodes

$$\Phi_{\mathrm{MIEC}} = U_{\mathrm{N}}(p_{\mathrm{CC}}, p_{\mathrm{CE}}) + \Phi_{\mathrm{CE}} - \eta_{\mathrm{Model}}. \tag{5.22}$$

Thus, only the movement of the oxygen ions is considered in the MIEC, which can be modelled using Fick's first law

$$J_{\mathrm{bulk}} = -D^{\delta}\nabla c_{\mathrm{O^{2-}}} \quad \text{in } \Omega_{\mathrm{MIEC}}. \tag{5.23}$$

The chemical diffusion coefficient D^{δ} depends on temperature and partial pressure, just like k^{δ}. In analogy to k^{δ}, a constant D^{δ} is used in the Rüger model and the early version of ParCell3D. In the current version, D^{δ} is partial pressure dependent, although the difference is extremely small (see Section 5.5.2). This is considered by

$$D^{\delta} = D^{\delta}_{\mathrm{ref}} \cdot (p\mathrm{O}_2/p\mathrm{O}_{2,\mathrm{ref}})^{\alpha_D}, \tag{5.24}$$

whereas reference values for D^{δ} and $p\mathrm{O}_2$ are used, and α_D is the slope of D^{δ} with varying $p\mathrm{O}_2$. However, the diffusion equation for the MIEC phase is therefore

$$\nabla \cdot J_{\mathrm{bulk}} = 0 \quad \text{in } \Omega_{\mathrm{MIEC}}, \tag{5.25}$$

with boundary conditions on interface Γ_{MP} given by Eq. 5.20b and the boundary condition between electrolyte and MIEC is given by Eq. 5.29b described in Section 5.2.3.

Charge Transfer

At the interface between MIEC and electrolyte (Γ_{ME}), oxygen ions are exchanged. The local charge transfer voltage

$$\eta_{\mathrm{ct}} = U_{\mathrm{N,ME}} - (\Phi_{\mathrm{MIEC}} - \Phi_{\mathrm{E,\,ME}}) \tag{5.26}$$

with

$$U_{\mathrm{N,ME}}(p_{\mathrm{M,ME}}, p_{\mathrm{CE}}) = \frac{RT}{2F}\log\sqrt{\frac{p_{\mathrm{M,\,ME}}}{p_{\mathrm{CE}}}} \tag{5.27}$$

is the driving force for this process. This charge transfer voltage depends on the local difference between the equilibrium reduction potential ($U_{\mathrm{N,ME}}$) and the difference between the the electrical potential of MIEC (Φ_{MIEC}) and electrolyte ($\Phi_{\mathrm{E,\,ME}}$) at the interface. A constant oxygen partial pressure ($p_{CE} = p_{El,IF}$) is assumed within the electrolyte which has to be defined, whereas in contrast the oxygen partial pressure $p_{M,\,ME}$ at the interface with the MIEC is being calculated and corresponds to the partial pressure with which the MIEC material would be in equilibrium due to the local oxygen ion concentration $c_{O,IF}$ at the interface [194, 196]. With this voltage the local oxygen ion exchange current density can be determined by

$$I_{ct} = \frac{\eta_{ct}}{ASR_{ct}}. \tag{5.28}$$

The charge transfer is modelled by the flux of oxygen ions at the interface

$$n \cdot J_{bulk} = \frac{I_{ct}}{2F} \quad \text{on } \Gamma_{ME}, \tag{5.29a}$$

$$n \cdot J_{elyt} = I_{ct} \quad \text{on } \Gamma_{EM}. \tag{5.29b}$$

The area specific charge transfer resistance ASR_{ct} is the only parameter of this model for which no values can be found in literature. If a LSCF/CGO material system is considered, a close to lossless charge transfer can be assumed [196]. The charge transfer resistance for other material combinations like LSCF/YSZ can be substantially higher, and even dominate the cathode resistance e.g. due to insulating secondary phases (lanthanides and zirconates). However, as long as no insulating secondary phase hinders the charge transfer at any of the interfaces, the ASR_{ct} can be neglected [204]. Therefore, a very small value of 10^{-4} Ωcm^2 for ASR_{ct} is used for the calculations, but using higher values is also possible.

Ionic Current in the Electrolyte

Electrical and ionic properties of materials are, just like bulk diffusion, described by a Poisson equation. A constant chemical potential determined by p_{CE} is assumed in the electrolyte, since the oxygen ion concentration is more or less independent of the oxygen partial pressure and because an ideal reversible counter electrode is used. The diffusion of oxygen ions in the electrolyte is thus modelled by a gradient of the electrical potential

$$J_{elyt} = -\sigma_{elyt} \nabla \Phi_{elyt} \quad \text{in } \Omega_{elyt}. \tag{5.30}$$

In this case the ionic conductivity of the electrolyte σ_{elyt} has a significant influence. Hence the equation for the potential in the electrolyte are

$$\nabla \cdot J_{elyt} = 0 \quad \text{in } \Omega_{elyt}, \tag{5.31a}$$

$$n \cdot J_{elyt} = 0 \quad \text{on } \Gamma_{EP}, \tag{5.31b}$$

$$\Phi_{E} = 0 \quad \text{on } \Gamma_{elyt}, \tag{5.31c}$$

where Γ_{elyt} is the surface underneath of the electrolyte, on which the potential is set to a reference value of zero. A further condition on the interface to the MIEC has been given in Eq. 5.29b (Section 5.2.3).

Counter Electrode

As already mentioned, the model considers an ideal reversible counter electrode with a constant potential. Thus it follows that

$$\begin{aligned} \Phi_{E,CE} &= \Phi_{CE} \quad \text{on } \Gamma_{E,CE}, \\ p_{E,CE} &= p_{CE} \quad \text{on } \Gamma_{E,CE}, \end{aligned} \tag{5.32}$$

whereas $\Phi_{CE} = 0$ in Ω_{CE} is set in the model.

Other Boundary Conditions

On the lateral boundaries of the considered volume (the computational domain), a no flux condition (insulation) is imposed

$$\begin{aligned} n \cdot J_{O_2} &= 0 \quad \text{on } \Gamma_0, \\ n \cdot J_{bulk} &= 0 \quad \text{on } \Gamma_0, \\ n \cdot J_{elyt} &= 0 \quad \text{on } \Gamma_0, \end{aligned} \tag{5.33}$$

where Γ_0 are the four lateral boundaries as depicted in Fig. 5.5.

5.2.4. Numerical Method

In this section details are given on the method used in ParCell3D to solve the problem described above, based on Ref. [119]. The use of a standard *iterative* solver from commercially-available software would not solve this ill-conditioned problem due to the specific properties of the equations and the coupling between them. It would be possible with a *direct* solver usually available in commercial software, but it requires a huge amount of memory to solve problem as complex as the one considered in this work. For these reasons ParCell3D was developed by T. Carraro (University of Heidelberg) with a parallel iterative solver, as described in the following.

Discretization and Solver

The finite element method is used to numerically discretize and solve the system of PDEs that describes the complete coupled problem in the C++ program ParCell3D. This software was specifically developed to solve this problem. For the discretization, the program relies on the FEM library deal.II [176]. The main parts of the program are: (a) a grid generator which takes the voxel data from the segmentation 3D image data and creates a mesh for the calculation, (b) the building of the system with the imposition of the fluxes (5.20) and (5.29) at the boundaries between different phases and (c) a parallel solver.

The mesh is created by setting subregions of the same phases, in which a continuous finite element ansatz is considered. Each voxel of the reconstruction is typically one finite element, but naturally it can also be refined or coarsened. Then the interfaces between different materials are identified. During the system assembling the boundary conditions (5.20) and (5.29) are used to couple the different subregions. This will result in a stiff nonlinear system of equations. The nonlinearity is solved by an exact Newton method, where the linear solver is parallelized.

The problem is quite difficult to solve due to the stiff coupling between the pore phase and the MIEC, regardless of the method used to discretize the problem. This can be evidenced by an eigenvalue decomposition of the system matrix. Because of the high condition number of the matrix and the peculiar clustering of the eigenvalues, an iterative solver needs a special preconditioner to converge (i.e. a numerical method which creates an iterative matrix with much better properties). The actual version of ParCell3D uses a GMRES solver [205] preconditioned by a domain decomposition method [206]. In the implemented domain decomposition method, the considered volume is divided into subregions. Each part of the volume is solved in parallel using a GMRES solver with an ILU [205] preconditioner.

As a simple example, the computational grid of one volume from cathode type 1 has more than 18 million voxels, which correspond to 23 million degrees of freedom using trilinear finite elements. It was found that when performing the simulations with up to 144 GB RAM on the high-performance computer HP XC3000 (short: hc3), the simulation of structures consisting of up to 25 million voxels is possible. The hc3 is housed at the information technology centre of the KIT (Steinbuch Centre of Computing, SCC).

Boundary Conditions and Output

As mentioned above, the overpotential η_{Model} of the model (cf. Eq. 5.22) has to be defined before calculation, as well as the potential at the bottom surface of the electrolyte layer (boundary condition on Γ_{elyt}) and the oxygen concentration in the pore phase at the top of the microstructure (boundary condition on Γ_{CC}, cf. Fig. 5.5b). These conditions induce a current through the microstructure.

The solution of this problem are space-resolved values such as the oxygen concentration and flux in the pore phase, the concentration of oxygen ions in the bulk phase and the potential in the electrolyte. The current is calculated by integrating the flux. For calculating the ASR_{Model} the total current I of the structure is calculated at Γ_{CC}. Integrating the current density inside the MIEC material at different heights of the cathode, a current profile can be calculated which shows how much load is already incorporated in the mixed conducting material. The calculated current profile is used to define the penetration depth l_δ (cf. Eq. (5.5), see also e.g. Section 5.5.2, Fig. 5.19).

However, the computational domain of this FEM model is based on voxel data, which one should bear in mind when comparing its results with e.g. homogenized or 1D models. Since the FEM model relies on the voxel data, the cubic grid overestimates the surface area density a compared to a more accurate calculation, e.g. by means of the marching cubes method described in Section 4.7.3. Although the marching cubes approximation is more precise, the surface area density from the voxel data has to be used in homogenized models, if comparing the results from ParCell3D. A direct comparison between the 3D calculation from voxel data and homogenized models utilizing the same data is thereby possible. A further improvement in the 3D calculation will be possible if the MC algorithm is included in the FEM calculation. This is part of ongoing work.

5.3. Influence of Material on Performance

The cathode performance predominantly depends on the material composition and microstructure of the porous cathodes, as already discussed. This section investigates the influence of material composition on cathode performance. In ParCell3D the material composition is represented by the material specific parameters k^δ and D^δ together with the oxygen ion concentration at equilibrium $c_{\mathrm{O^{2-},eq}}$. These parameters are all dependent on temperature and oxygen partial pressure and therefore have to be determined from literature for all operating conditions which are to be simulated. The oxygen ion concentration is taken as described in Table 5.1. For k^δ and D^δ, however, a large scatter can be found in the published values, even for (nominally) the same composition (cf. Fig. 5.7).

The cathodes analysed in this thesis are $La_{0.58}Sr_{0.4}Co_{0.2}Fe_{0.8}O_{3-\delta}$ from FZJ. Since very little is reported in literature for this specific composition, also k^δ and D^δ values for $La_{0.6}Sr_{0.4}Co_{0.2}Fe_{0.8}O_{3-\delta}$ are investigated. The slightly higher amount of lanthanum might lead to slightly different material parameters, but this influence is negligible compared to the scatter found between different groups. However, in this section the same microstructure (reconstruction of cathode 2_{AP}) is used for all simulations with different material parameters k^δ and D^δ, while the impact on the ASR_{cat} is studied. Figure 5.7 compares k^δ and D^δ values as published by Leonide [2], Søgaard [196], Bouwmeester [199] and Ried [207] at temperatures between 600 °C and 800 °C (note that the values from [2] for D^{δ} had to be corrected due to an incorrect assumption in Ref. [2]; see Section 5.2.1). Both parameters increase with increasing temperature, whereas some of the literature data show Arrhenius type behaviour (linear course of the data in this plot) over a broad temperature range (Refs. [2, 196]). The activation energies E_a are between 1.26 eV [2] and 1.37 eV [196] for D^{δ}, while for k^{δ} the activation energies are difficult to compare, since most of the literature data do not consistently show Arrhenius behaviour over the entire temperature range (except for the values of [2] with $E_a = 1.05$ eV). However, it can be seen that the scatter for k^{δ} and D^{δ} can range up to one order of magnitude.

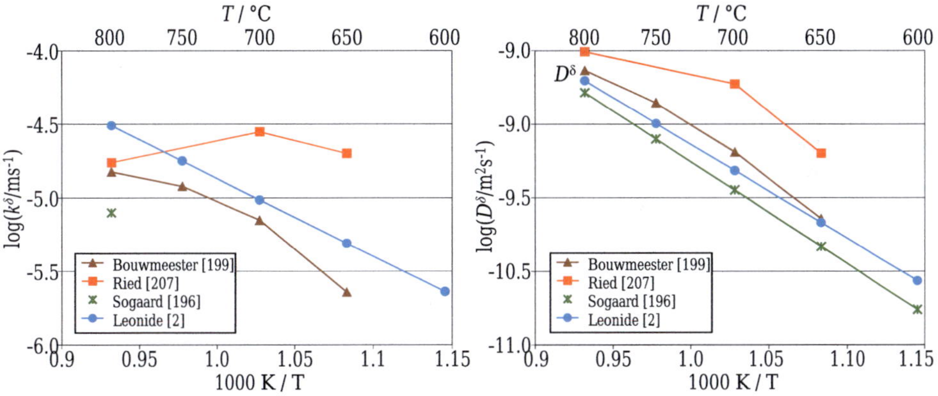

Figure 5.7.: Summary of values for (a) the surface exchange coefficient k^{δ} and (b) the bulk diffusion coefficient D^{δ} in air for $La_{0.58/0.6}Sr_{0.4}Co_{0.2}Fe_{0.8}O_{3-\delta}$ at different temperatures as published in Literature [2, 196, 199, 207].

For a given microstructure (here, cathode 2_{AP}), the penetration depth l_{δ} also shows a certain scatter due to the different k^{δ} and D^{δ} values, as shown in Fig. 5.8. The lowest l_{δ} is calculated from the k^{δ} and D^{δ} values published by Leonide [2]. With increasing temperature, l_{δ} slightly increases, due to the higher activation energy for D^{δ} ($E_a = 1.26$ eV) compared to k^{δ} ($E_a = 1.05$ eV). Using the values from Ried [207] leads to an even faster increase of l_{δ} with temperature. In contrast, if using the values from Søgaard [196] (with D^{δ} as extrapolated in [26]), l_{δ} is decreasing slightly from 2.68 to 2.81 µm with increasing temperature, whilst showing the overall greatest penetration depth. Note, that all simulations shown here are for a pO_2 of 0.21 bar.

The influence of the different material parameters on the cathode performance can be seen in Fig. 5.9, where the temperature dependent ASR_{cat} is depicted as calculated with the k^{δ} and D^{δ} values shown in Fig. 5.7. It can be seen that the ASR_{cat} increases with increasing temperature, since both, the surface exchange and oxygen ion diffusion in the MIEC material slow down with decreasing temperature.

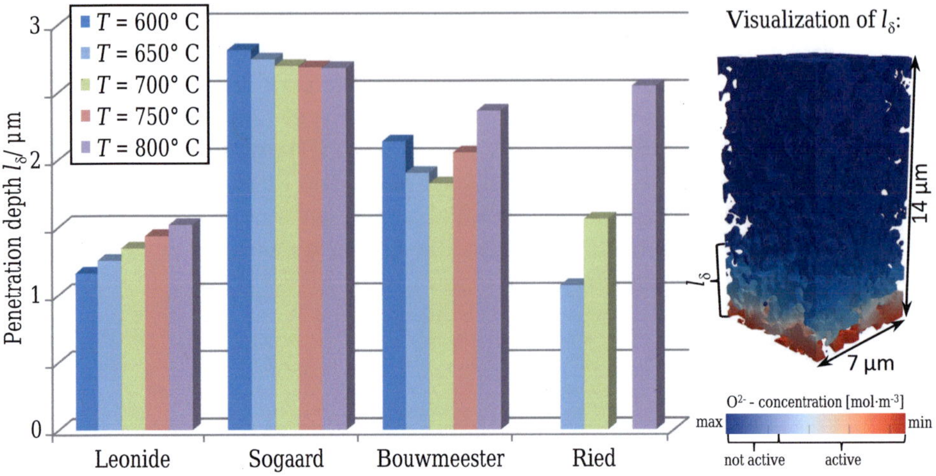

Figure 5.8.: Penetration depth l_δ at different temperatures, calculated for k^δ and D^δ values published by Leonide [2], Søgaard [196] (using D^δ as extrapolated in [26]), Bouwmeester [199] and Ried [207]. The microstructural parameters are from cathode 2_{AP}.

Naturally, the simulation using the lowest values for k^δ and D^δ shows the highest ASR_{cat} values (Ref. [196]), meanwhile the lowest resistance is calculated from the highest material parameters (Ref. [207]). For the sake of comparison, measurement data from Ref. [208] for a state of the art cathode from FZJ are also plotted in Fig. 5.9, positioned in the middle of the simulated results. However, it should be noted that the k^δ and D^δ values from Leonide [2] are obtained from these measurements. It is thus not surprising that the simulated values based on the material parameters from Leonide show the same activation energy and that the simulated ASR_{cat} values are very close to the measured resistances (less than 10 % difference). Actually, the only reason why the measured and simulated values are not even closer to each other are that the values for the thermodynamic factor γ and of the microstructural parameters a, ε and τ_{LSCF} used in [2] to calculate k^δ and D^δ were not precisely known at that time. However, the values for the ASR_{cat} also spread up to one order of magnitude, depending on the temperature. It becomes obvious that the material parameters have a large influence on the calculated performance, and that it is crucial for the accuracy of the results to know these parameters to a high degree of certainty. The next section therefore presents a method for the accurate determination of k^δ and D^δ.

A more analytical approach for the quantification of the influence of k^δ and D^δ on the calculated ASR_{cat} with ParCell3D was presented in Ref. [177], on which the following paragraph is based. A sensitivity factor with respect to different model parameters (e.g. k^δ and D^δ) is defined and calculated, using the k^δ and D^δ values from Ref. [2] and the microstructure of cathode type 1. Here the figure of interest is the variation of the ASR_{cat}

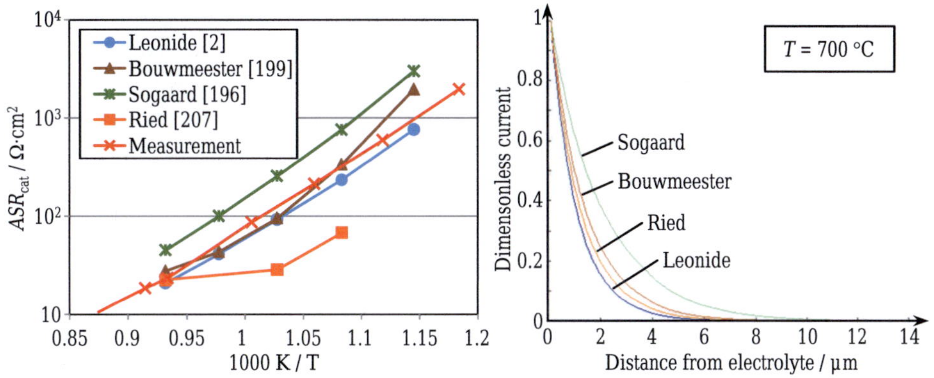

Figure 5.9.: (a) The area specific resistance and (b) corresponding current profiles at $T = 700$°C calculated with ParCell3D for different k^{δ} and D^{δ} values for LSCF from literature (cf. Fig. 5.7).

(denoted as δASR_{cat} in the following), in dependency of the variation of a model parameter p (δp is the variation of the parameter p), which can be expressed as

$$\delta ASR_{\text{cat}} \approx \frac{\partial ASR_{\text{cat}}}{\partial p} \delta p. \tag{5.34}$$

As the model parameters vary by several orders of magnitude, the parameters must be first normalized by writing the model parameter p as multiple of a reference value p_{ref}, by multiplying it with a normalization factor $\tilde{p}$

$$p = p_{\text{ref}} \tilde{p}. \tag{5.35}$$

The derivative of the quantity of interest (which is the ASR_{cat} here) with respect to the normalized parameter $\tilde{p}$ is the sensitivity factor

$$SF = \frac{\partial ASR_{\text{cat}}}{\partial \tilde{p}} = \frac{\partial ASR_{\text{cat}}}{\partial p} \frac{p}{\tilde{p}} = \frac{\partial ASR_{\text{cat}}}{\partial p} p_{\text{ref}}. \tag{5.36}$$

The advantage of using the SF is that it allows a comparison of the sensitivities independently from the different scales of the parameters. It is more intuitive, however, to use the relative sensitivity factor (RSF), which is the SF divided by the value ASR_{cat} at the considered temperature (cf. Table 5.2). By using the RSF, the influence of k^{δ} and D^{δ} on the ASR_{cat} can be easily estimated. A variation of x% of the model parameter p leads to a variation of $x \cdot \text{RSF}(p)$ percent for ASR_{cat}, e.g. at 600 °C a variation of 10 % of D^{δ} will change the ASR_{cat} by 4.83 %. The temperature-dependent relative sensitivity factors with respect to k^{δ}, D^{δ} and D_{gas} are listed in Table 5.2. It can be seen that the RSFs for k^{δ} and D^{δ} are pretty similar, while the RSF(D_{gas}) values are much smaller. This is not surprising and proves that the gas diffusion in the studied case has an extremely small influence and can be neglected. The fact that RSF(k^δ) and RSF(D^δ) are close to each other illustrates that LSCF at the studied conditions is controlled by both k^δ and D^δ. However, even if the sensitivity factors allow quantitative calculation of the importance of k^δ and D^δ (and other model parameters, see Ref. [177]) in regard to the calculated ASR_{cat}, the large scatter in the k^δ and D^δ values published in literature is still unexplained (see Section 5.4 for more information).

Table 5.2.: Relative sensitivity factor (RSF=SF/ASR_{cat}) of ParCell3D.

T / °C	ASR_{cat} / $\Omega\mathrm{cm}^2$	RSF(k^δ)	RSF(D^δ)	RSF(D_{gas})
600	0.614	−0.483	−0.513	$7.54 \cdot 10^{-10}$
650	0.188	−0.484	−0.509	$2.82 \cdot 10^{-9}$
700	0.074	−0.483	−0.505	$1.43 \cdot 10^{-8}$
750	0.033	−0.480	−0.498	$3.73 \cdot 10^{-8}$
800	0.019	−0.477	−0.492	$8.58 \cdot 10^{-8}$

A possible explanation arises, if comparing k^δ and D^δ values published for different cathode materials. The material parameters for three different MIEC materials are compared here (LSCF: from Leonide [2]; LSC: from De Souza *et al.* [209] and interpolated values thereof as in Ref. [26]; BSCF: from Ried *et al.* [207] and Girdauskaite *et al.* [210] as interpolated in Ref. [26]). Values for $c_{\mathrm{O^{2-},eq}}$ for all three materials are taken from Ref. [26]. A comparison of the parameters k^δ and D^δ for different temperatures is given in Fig. 5.10.

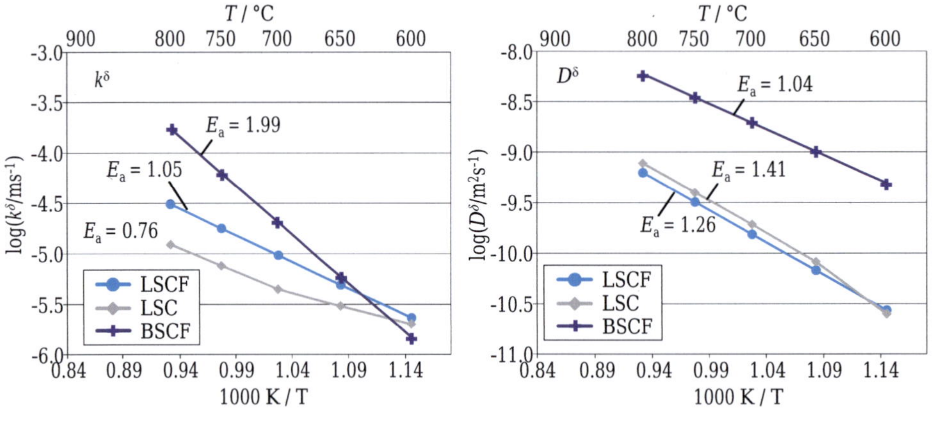

Figure 5.10.: Material parameters (a) surface exchange coefficient k^δ and (b) bulk diffusion coefficient D^δ in air for different MIEC materials between 600 °C and 800 °C (values for LSCF from Ref. [2]; values for LSC and BSCF from Ref. [26], interpolated from Refs. [207, 209, 210]).

At 800 °C the surface exchange coefficient of BSCF is much larger than for LSCF, and even more than for LSC, which shows the lowest value for k^{δ} given here. If looking at the activation energies of the three materials, BSCF shows the highest activation energy (1.99 eV), while LSC shows the lowest (0.76 eV). Thus, with decreasing temperature the values for all three materials get closer, until at 600 °C the k^{δ} is nearly the same for all three materials. In contrast, the diffusion coefficient D^{δ} shows much more similar activation energies of between 1.04 eV (BSCF) and 1.41 eV (LSC) for the three materials. BSCF has by far the greatest D^{δ} values over the whole investigated temperature range, while LSC is slightly above the values for LSCF.

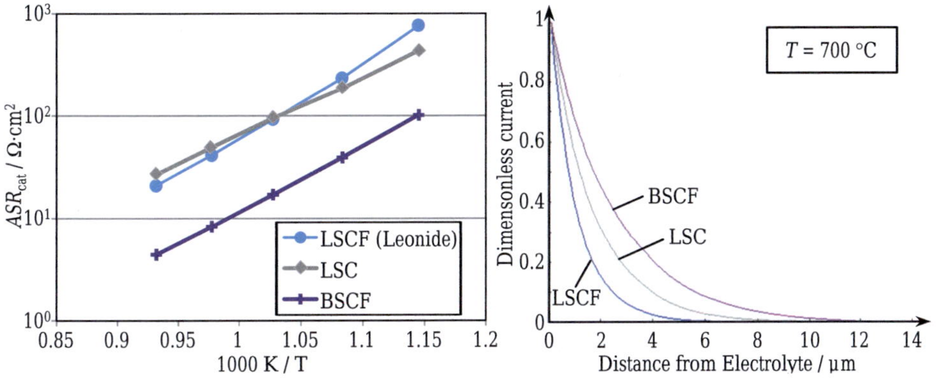

Figure 5.11.: (a) The area specific resistance and (b) corresponding current profiles at $T = 700$°C calculated with ParCell3D for different k^{δ} and D^{δ} values for LSCF, LSC and BSCF from literature (cf. Fig. 5.10).

Similarly to the study above for different LSCF k^{δ} and D^{δ} values, the ASR_{cat} is calculated using the reconstruction of cathode 2_{AP} as computational domain. The results shown in Fig. 5.11 illustrate that above 700 °C the ASR_{cat} of the LSCF cathode is slightly lower compared to the LSC cathode, which changes below 700 °C, where the ASR_{cat} for LSC is lower than for LSCF due to the different activation energies of k^{δ}. However, due to the large scatter of the material parameters published by different groups, it is not certain which material is better, as the ASR_{cat} values are very close to each other.

Interestingly, the values for BSCF are much lower (between 3.3 and 6.5 times lower) compared to LSC or LSCF. If looking at these results, BSCF seems to be a much better candidate for SOFC cathodes. In fact, BSCF was presented as a very promising candidate for intermediate-temperature SOFCs some years ago. But BSCF is not used in todays state of the art SOFCs (at least if operated in ambient air) due to its pronounced instability in the presence of even low amounts of carbon dioxide, which leads to the formation of carbonates and, thereby, the significant lowering of k^{δ} over time [3,211,212]. However, this clearly underlines that the material parameters depend on the stability of the sample and especially on the stage of degradation, as this naturally changes k^{δ} and D^{δ}. Interestingly, very little has been reported to date about degradation and the time-dependent change of the material parameters k^{δ} and D^{δ}. This will be discussed in the next section, where a method is presented that can determine k^{δ} and D^{δ} to a high degree of accuracy. Moreover, this method is able to track the time-dependent course of the parameters during operation.

5.4. Calculation of Material Parameters k^{δ} and D^{δ}

The large scatter of values for the surface exchange coefficient k^{δ} and the chemical diffusion coefficient D^{δ} published in literature (cf. Fig. 5.7), even for the same materials, clearly shows the need for an accurate method of determining these parameters. As mentioned in Section 2.4.1, k and D are typically assessed from dense bulk samples (i.e. from chemical diffusion experiments (k^{δ}, D^{δ}), electrical conductance experiments (k^{Q}, D^{Q}) or tracer experiments (k^{*}, D^{*})).

However, these experiments struggle with several problems [39]. The dense samples themselves are a large source of error. They are typically fabricated by sintering at higher temperatures than porous structures, which might influence the chemical composition (thus changing k and D) or at least the grain size of the sample (thus changing the grain-to-grain boundary ratio, possibly influencing both k and D). In general, the sample specific conditions might have changed due to the thermal history of the sample prior to measurement – i.e. the fabrication process of the sample as well as eventual prior unrelated treatments or measurements. The sample specific conditions are prone to change during the measurement itself, i.e. changes occur if the experiment runs for prolonged periods or at different temperatures. That might be one reason for the scatter in literature values, but since no information is available about the time dependent behaviour of k and D in dense bulk samples, this remains speculation. A more comprehensive discussion of possible sources of error can be found in [39].

A method capable of determining k and D of a porous MIEC cathode as it is being used in a state of the art SOFCs is highly valuable. Such a method is presented in this section, which is based on Ref. [39]. This approach is capable of both, determining the material parameters of relevant electrodes with a high accuracy, and of tracking degradation phenomena correlated to microstructural changes (e.g. coarsening of the structure and the associated reduction in the surface area density) or material changes (e.g. demixing or decomposition, etc.). Please note that in the following, the chemical diffusion coefficient D^{δ} and the associated surface coefficient k^{δ} are used, but the different types of material coefficients can be converted into each other (cf. Section 2.4.1).

5.4.1. Method and Parameters

The presented method, reported e.g. in Refs. [2,27], uses EIS measurements in combination with microstructure modelling and microstructural information (obtained e.g. by FIB tomography) to determine the material parameters k^δ and D^δ. From the measured impedance spectra, the area specific cathode resistance ($ASR_{cat} \equiv R_{chem}$) and the characteristic time constant (t_{chem}) of the porous MIEC cathode can be obtained by a CNLS fit (complex nonlinear least squares; cf. Refs. [2,213]) using the equivalent circuit model (ECM) developed in Ref. [2]. In this ECM, the cathode is modelled using a Gerischer Element as in the ALS model [60]. Thus, the following equations derived from the ALS model (cf. Section 5.1) can be used to calculate the surface exchange coefficient k^δ and the chemical diffusion coefficient D^δ

$$k^\delta = \frac{c_{O^{2-}}(1-\varepsilon)}{a c_{mc} t_{chem}}, \tag{5.37}$$

$$D^\delta = \left(\frac{RT}{4F^2}\right)^2 \frac{\tau_{LSCF}\gamma_o^2 t_{chem}}{(1-\varepsilon)^2 c_{mc} c_{O^{2-}} R_{chem}^2}, \tag{5.38}$$

provided that the parameters ε, a, τ_{LSCF}, $c_{O^{2-}}$, c_{mc} and γ_o are known. Please note that these equations are written in terms of the oxygen ion diffusivity (k^δ and D^δ) and thermodynamic factor γ_o, whereas Eqs. (5.3) and (5.4) from the ALS model are written in terms of oxygen vacancies as mobile species. Appendix B explains the transformations used to derive Eqs. (5.37) and (5.38).

However, in the framework of this thesis (and for the first time) actual values for ε, a and τ_{LSCF} were used to calculate the specific k^δ and D^δ values [39]. In Refs. [2,27] these structural parameters were only roughly estimated, which had led to errors or at least significant uncertainties. From Eqs. (5.37) and (5.38) it can be seen that an error in the assumed surface area density a will only affect the value for k^δ, while an error in the value of the tortuosity τ_{LSCF} will only affect D^δ. In contrast, the porosity fraction ε affects both parameters, with a more pronounced effect on D^δ.

Values for oxygen lattice sites c_{mc} and the O^{2-} concentration $c_{O^{2-}}$ can be found in literature (e.g. [26,196,199,202]) for $La_{0.58/0.6}Sr_{0.4}Co_{0.2}Fe_{0.8}O_{3-\delta}$. The two parameters can also be linked by the oxygen nonstoichiometry (fraction of oxygen vacancies) δ

$$c_{mc} = \frac{3}{3-\delta} c_{O^{2-}}. \tag{5.39}$$

Values from literature for $c_{O^{2-}}$ as a function of the partial pressure for different temperatures are summarized in Ref. [26], where a functional dependency between $c_{O^{2-}}$ and $\log(pO_2)$ is given in the pO_2 range of between approximately 10^{-4} and 1 bar (cf. Eq. (5.16) and the explanation above the equation). However, since the pO_2 dependent values are not perfectly linear over the whole pO_2 range, a more accurate fit of the parameters can be obtained if the values are only fitted in the pO_2 range relevant for SOFC cathodes (between 10^{-2} and 1 bar). The values $C_1(T)$ and $C_2(T)$ obtained between 10^{-2} and 1 bar are listed in Table 5.3. With these values, the concentration of oxygen ions can be calculated according to Eq. (5.16). The obtained values for $c_{O^{2-}}$ and c_{mc} are also listed in Table 5.3 together with values for δ, taken from Ref. [199]. These values were used to calculate k^δ and D^δ in this section.

Table 5.3.: Summary of material parameters used to calculate k^δ and D^δ in this section.

T / °C	$C_1(T)$	$C_2(T)$	$c_{O^{2-}}$ / mol · m^{-3}	c_{mc} / mol · m^{-3}	δ
600	267	85 317	85 136	85 525	0.0137
650	396	84 891	84 623	85 106	0.0171
700	525	84 465	84 109	84 740	0.0223
750	653	84 039	83 596	84 424	0.0294
800	782	83 613	83 083	84 155	0.0382
850	910	83 187	82 570	83 926	0.0485
900	1039	82 761	82 057	83 732	0.0600

Temperature dependent values for the thermodynamic enhancement factor γ_o can also be found in literature (e.g. [66,214,215]), or they can be determined from $c_{O^{2-}}$ or δ, respectively. Following Ref. [26], γ_o can be calculated by using $C_1(T)$ and $C_2(T)$ from Table 5.3 with the following equation

$$\gamma_o(T, p_{O_2}) = \frac{1}{2}\ln(10) \cdot \left(\log(p_{O_2}) + \frac{C_1(T)}{C_2(T)}\right). \tag{5.40}$$

The values for γ_o calculated in this manner are given in Table 5.4, where other literature values are also listed [66, 214, 215] for comparison. For example, den Otter determined the oxygen nonstoichiometry δ versus the partial pressure by thermogravimetric measurements on $La_{0.6}Sr_{0.4}Co_{0.2}Fe_{0.8}O_{3-\delta}$ bulk samples, from which he calculated the thermodynamic factor. These values were fitted with an exponential function in Refs. [2, 39] and used to determine k^δ and D^δ. However, in order to be consistent and use the same data for $c_{O^{2-}}$ and γ_o, the values calculated with $C_1(T)$ and $C_2(T)$ from Table 5.3 are used to determine the material parameters in this section.

Although the determination of these parameters is beyond the focus of this work, it should be noted that the experimental determination of δ, and hence γ_o, is rather complex and could be subject to error, especially at lower temperatures. Also, some conflicting results have been published in literature. For example, in [214] γ_o is approximately constant in the pO_2 range from 1 to 10^{-2} bar, while in [66] γ_o decreased by almost an order of magnitude as the pO_2 decreased from 1 to $10^{-2.5}$ bar. However, at the conditions applied for the cells investigated here (air, $pO_2 = 0.21$ bar), the literature values are in good agreement (at $T = 750$ °C: value used here is $\gamma_o = 147$, while the interpolated value from [214] is ≈ 187), see Table 5.4. It is worth noting that γ_o (and hence also δ) influences, almost exclusively, the value of D^δ, while k^δ is only very slightly affected (cf. Eqs. (5.37) and (5.38)). Nevertheless, the uncertainty of the thermodynamic factor might be the largest source of error for this method. Table 5.5 shows the ranges of all parameters, as well as their mean values used for the calculations described below.

Table 5.4.: Thermodynamic factor γ_o as determined using $C_1(T)$ and $C_2(T)$ from Table 5.3 and used for calculating k^δ and D^δ. Also, literature values from [66,214,215] for comparison.

T / °C	Using $C_{1,2}(T)$ from Table 5.3	Ref. [214]	Interpolated from [214]	Ref. [66]	Calculated from δ as in Ref. [215]
600	366	409	409	N/A	2281
650	246	N/A	306	N/A	877
700	185	240	236	N/A	418
750	147	N/A	187	N/A	260
800	122	142	151	~202	187
850	104	N/A	125	N/A	149
900	91	111	104	N/A	119

Table 5.5.: Microstructural parameters (calculated from cathode 2_{AP}) and thermodynamic parameters used in Eqs. (5.37) and (5.38) to calculate k^δ and D^δ from R_{chem} and t_{chem}.

Parameter	LSCF	Source	Mean
Porosity ε / %	$42.4-46.8(44.6\pm2.2)$	FIB tomo.	44.6
Tortuosity LSCF τ_{LSCF}	$1.51-1.85(1.68\pm0.17)$	FIB tomo.	1.68
Surface area density a / µm^{-1}	$2.59-3.17(2.88\pm0.29)$	FIB tomo.	2.88
Ox. latt. sites c_{mc} / mol · m^{-3}	$83732-85525$	Eq. (5.39)	84713
O^{2-} conc. $c_{O^{2-}}$ / mol · m^{-3}	$82057-85136$	Eq. (5.16)	83919
Thermodyn. factor γ_o	$91-366$	Eq. (5.40)	147

5.4.2. Calculating the Material Parameters of Type 2 Cathodes

The material parameters k^δ and D^δ of cathode type 2 (standard LSCF cathode from FZJ) will be determined here. Measurement data from Ref. [27] are used, where three ASCs were electrochemically characterised over an operation time of over 1000 h at temperatures of 600, 750 and 900 °C. There, the ohmic and polarization losses of electrolyte, anode and cathode were identified and separated by high-resolution impedance studies using the ECM described in Ref. [37]. Moreover, the time dependent courses of the cathode polarization resistance R_{chem} and time constant t_{chem} were calculated (see Refs. [27,38] for more details), which are now used to determine k^{δ} and D^{δ}. The microstructural characterisation was realized on exactly these measured cells.

The time-dependent courses of both R_{chem} and t_{chem} are shown in Fig. 5.12, where a clear dependency to the operating temperature can be observed. At $T = 600\,°\mathrm{C}$, both parameters increase exponentially over the whole operating time, thus indicating a pronounced, continuous change of k^{δ} and D^{δ} over time.

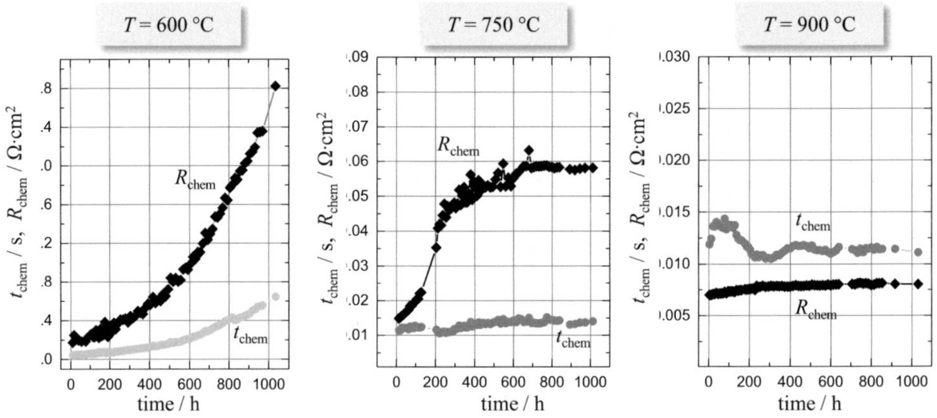

Figure 5.12.: Time-dependent courses of R_{chem} and t_{chem} associated with the Gerisher element and extracted from the impedance data by a CNLS fit. The parameters are shown for $T = 600\,°\mathrm{C}, 750\,°\mathrm{C}$ and $900\,°\mathrm{C}$ as calculated in [39].

This changes at higher temperatures. At $T = 750\,°\mathrm{C}$, a strong nonlinear increase for R_{chem} in the first 300 h is followed by a fairly constant plateau. In contrast, t_{chem} remains approximately constant. It is worth noting, that the increase of R_{chem} over time at $T = 750\,°\mathrm{C}$ was found to be reversible when the sample was kept at 900 °C for 160 h in a previous study [27]. This experiment was not performed at 600 °C and thus reversibility can only be assumed [39]. Nevertheless, the reversibility of the degradation is a strong indication that the changes in R_{chem} and t_{chem} are not set by microstructure, but rather by the material.

At $T = 900\,°\mathrm{C}$ R_{chem} increases only marginally, while t_{chem} shows a transient oscillation behavior. This was found to be due to measurement uncertainties rather than due to a change in the electrode characteristics, because the cathode polarization resistance is extremely low ($R_{\mathrm{pol}} = 9.72 \cdot 10^{-2}\,\Omega$) at 900 °C (for reasons of comparison: $R_{\mathrm{pol}} = 3.804\,\Omega$ at 600 °C and $R_{\mathrm{pol}} = 2.45 \cdot 10^{-1}\,\Omega$ at 750 °C).

Overall, the EIS analysis unambiguously reveals that the kinetics of the porous LSCF cathode change considerably with time at 600 °C over 1000 h, but less pronouncedly at

750 °C and remain nearly unaffected at 900 °C, see Fig. 5.12. This is a further indication that the changes in R_{chem} and t_{chem} are not set by microstructure, since such changes would be expected to be at their highest at $T = 900$ °C.

This assumption was already proven in Section 4.9, where the results of the microstructural analysis via FIB tomography of the three post-operation cathodes and an initial cathode of type 2 were compared. The three operated cells (2_{600}, 2_{750} and 2_{900}) showed almost no difference after operation of over 1000 h: ε differs by less than 2 %, τ_{LSCF} by less than 3 % and a by less than 3.5 % between all four cells (cf. Table 4.10). From the four cathodes, cathode 2_{900} shows the largest, but still negligible, differences. As the differences are so small, the values for a, ε and τ_{LSCF} from the not operated cathode 2_{AP} are used for the calculation of all k^δ and D^δ values. Nevertheless, to illustrate how a slightly different microstructure would affect k^δ and D^δ, error bars are calculated for an even larger scatter ($\varepsilon \pm 5$ %, $a \pm 10$ %, $\tau_{LSCF} \pm 10$ %) of the microstructural parameters than observed between the analysed structures. The trends of the material parameters are shown in Fig. 5.13, where the courses of R_{chem} and t_{chem} are displayed again in light grey to demonstrate the correlation between the parameters. Table 5.6 lists the values of k^δ and D^δ at the start of the long-term measurement ($t = t_{start}$), after approximately 300 h of operating time ($t_{\sim 300h}$) and at the end of the long-term measurement ($t = t_{end}$). Moreover, the penetration depth l_δ is estimated using Eq. (5.5) with k^δ and D^δ and the microstructural parameters from cathode 2_{AP} (cf. Table 5.5). Because the microstructural characteristics do not change between the different cells or over time, the ratio between k^δ and D^δ determines the differences between the values of l_δ exclusively. The results are listed in Table 5.6 and illustrated in Fig. 5.12.

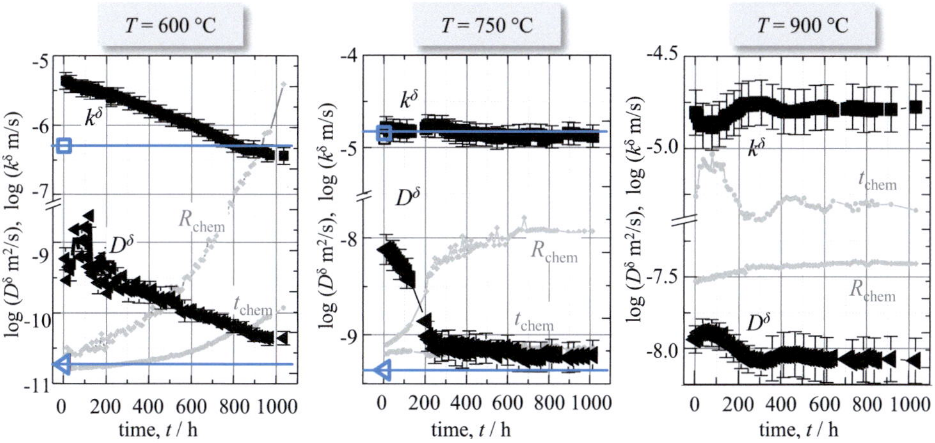

Figure 5.13.: Time-dependent courses of k^δ and D^δ as calculated from Eqs. (5.37) and (5.38) using R_{chem} and t_{chem} as published in [27], and the parameters listed in Table 5.5 (literature values in blue). The parameters are calculated for $T = 600$ °C, 750 °C and 900 °C.

Table 5.6.: Calculated values for k^δ and D^δ at $p_{O_2} = 0.21$ bar and corresponding penetration depth l_δ using Eq. (5.5).

Time t	$T = 600\,°C$		$T = 750\,°C$		$T = 900\,°C$	
	$\log k^\delta$	$\log D^\delta$	$\log k^\delta$	$\log D^\delta$	$\log k^\delta$	$\log D^\delta$
t_{start}	−5.36	−9.23	−4.79	−8.32	−4.81	−7.94
$t_{\sim 300h}$	−5.66	−9.68	−4.80	−9.32	−4.75	−8.09
t_{end}	−6.44	−10.36	−4.88	−9.41	−4.77	−8.09

Even if the interpretation of the results is not the main focus here (see Ref. [39] for more details), it can still be seen that the aging of the material is highly dependent on the operating temperature. At 600 °C, the strong increase of the polarization resistance can be attributed to both, k^δ and D^δ, as both parameters are more or less constantly decreasing. As this decrease is similar for both parameters, l_δ stays approximately constant at about 3.3 to 3.9 µm (cf. Fig. 5.14). For 750 °C the cathode polarization resistance changed nonlinearly, resulting in a rather constant value for k^δ but a distinct decrease for D^δ. The course of the chemical diffusion coefficient seems to be almost exclusively responsible for the strong increase in cathode polarization resistance (D^δ changed by one order of magnitude over 1000 h of operation). As a consequence of the increasing diffusion coefficient D^δ and constant surface exchange coefficient k^δ, the penetration depth l_δ decreases from 5.8 to 1.8 µm within 1000 h. At 900 °C, k^δ remained almost constant, while D^δ had only decreased marginally, caused by a slight increase of the cathode polarization resistance. However, due to the larger ratio between D^δ and k^δ at 900 °C, the penetration depth was much greater, with 9.2 to 7.2 µm (slightly decreasing in the first 200 h; cf. Fig. 5.14).

Values for k^δ and D^δ reported in literature for $La_{0.6}Sr_{0.4}Co_{0.2}Fe_{0.8}O_{3-\delta}$ are displayed with open (blue) symbols for $T = 600\,°C$ and 750 °C in Fig. 5.13 (no values for $T = 900\,°C$ in air have been published so far). It can be seen that the values calculated here are above the literature values, at least at the beginning of the measurements, while they do get progressively closer and closer over time. At 600 °C the k^δ value from literature (Ref. [26]; extrapolated value from Ref. [199]) corresponds to the value after approximately 820 h here, while at 750 °C the k^δ values are in good agreement (values at 750 °C from Ref. [199]). In contrast, D^δ is larger over the whole operation time at both temperatures, even if the values after 1000 h are become closer. Two plausible alternative hypotheses exist for the difference between the values determined here and the literature values. First, the values reported in Ref. [199], determined for $La_{0.6}Sr_{0.4}Co_{0.2}Fe_{0.8}O_{3-\delta}$ on bulk samples, were measured on a post-degradation sample. Second, the surface area a of the bulk sample in [199] was underestimated because the surface roughness of the cut and polished faces was not taken into account. Although no values at 900 °C in air have been published so far, the calculated k^δ and D^δ values are in good agreement with values recently published by Niedrig *et al.* [216], when extrapolating their values (calculated at lower oxygen partial pressures) to a pO_2 of 0.21 atm (applied in these studies). A more detailed discussion on this topic, including a discussion on the time-dependent behaviour of the calculated k^δ and D^δ values, can be found in Ref. [39].

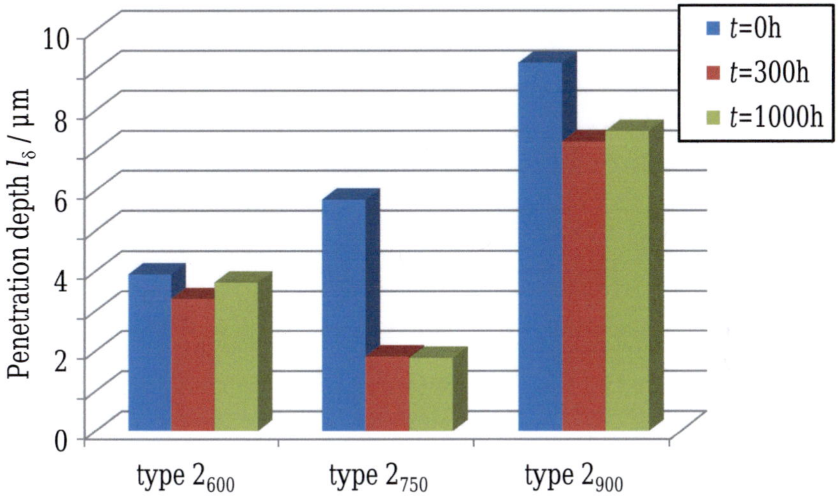

Figure 5.14.: Illustration of the penetration depth l_δ calculated with Eq. (5.5) from the k^δ and D^δ values as listed in Table 5.6 and the microstructural parameters from cathode 2_{AP} (cf. Table 5.5).

However, in this study it could be shown for the first time, that the degradation of the LSCF cathodes performance observed at different temperatures is almost exclusively caused by material characteristics, and not by a change in the LSCF microstructure. Moreover it was possible to track the courses of k^δ and D^δ over time and quantify the degree of degradation for these material parameters in dependence to the operating temperature for 600 °C, 750 °C and 900 °C.

5.4.3. Refining the Material Parameters from Leonide

As already mentioned, the method for determining k^δ and D^δ described above has also been used in Ref. [2] to calculate the material parameters for a cathode similar to type 2. However, in contrast to the calculation discussed above, the microstructural parameters were only roughly estimated in [2]. The largest source of error in these calculations was reported to be the uncertainty about the microstructural parameters. Hence, it shall be investigated here how the material parameters change when the microstructural parameters are estimated more accurately. Although the exact same cathode used in [2] is not reconstructed here, it can be assumed that the parameters calculated for cathode 2_{AP} are also representative for this cathode, since both cathodes are state of the art cathodes from FZJ.

Table 5.7.: Temperature dependent ASR_{cat} values (R_{chem}) and characteristic time constant (t_{chem}) of the state of the art LSCF cathode from FZJ, taken from Ref. [2]. Based on this data, the listed k^{δ} and D^{δ} values at $pO_2 = 0.21$ bar are calculated.

T / °C	R_{chem} / Ωcm^2	t_{chem} / s	D^{δ} / $m^2 s^{-1}$	k^{δ} / ms^{-1}
870	0.011	0.0039	$1.81 \cdot 10^{-9}$	$4.84 \cdot 10^{-5}$
820	0.019	0.0046	$8.54 \cdot 10^{-10}$	$4.12 \cdot 10^{-5}$
798	0.023	0.0052	$6.67 \cdot 10^{-10}$	$3.61 \cdot 10^{-5}$
721	0.086	0.0160	$2.27 \cdot 10^{-10}$	$1.18 \cdot 10^{-5}$
671	0.210	0.0310	$1.13 \cdot 10^{-10}$	$6.09 \cdot 10^{-6}$
621	0.599	0.0602	$5.06 \cdot 10^{-11}$	$3.14 \cdot 10^{-6}$
571	1.940	0.1385	$2.18 \cdot 10^{-11}$	$1.36 \cdot 10^{-6}$

For the calculations in Ref. [2], the microstructural parameters were estimated from 2D SEM images and the Rüger model [26] as $\varepsilon = 35\,\%$, $a = 3.32\,\mu m^{-1}$ and $\tau_{LSCF} = 1.38$, while the parameters from cathode 2_{AP} are determined from the 3D reconstruction as $\varepsilon = 44.6\,\%$, $a = 2.88\,\mu m^{-1}$ and $\tau_{LSCF} = 1.68$. The large porosity discrepancy shows, again, how hard it is especially to estimate the porosity fraction from SEM images (cf. Fig. 2.6). The thermodynamic parameters $c_{O^{2-}}$, c_{mc} and γ_o are taken as listed in Table 5.5. It should be noted that in Ref. [2] different material parameters as well as different values for $c_{O^{2-}}$, c_{mc} and γ_o were applied, which influences k^{δ} and D^{δ} as will be discussed below. The temperature dependent ASR_{cat} (R_{chem}) values and the corresponding characteristic time constants t_{chem} from Ref. [2] are listed in Table 5.7. These values are used to calculate k^{δ} and D^{δ} with the accurate microstructural parameters from cathode type 2 as described above. The newly calculated values are also listed in Table 5.7 and shown in Fig. 5.15 alongside the "old" values published in Ref. [2].

Interestingly, the old values (light blue symbols) and newly calculated *refined* values (dark blue symbols) for k^{δ} barely differed (less than 3 % difference). This is due to the fact that k^{δ} is determined by ε and a. The surface area density was overestimated in the rough estimation in Ref. [2], which lead to an underestimation of k^{δ}. In contrast, ε was underestimated, which lead to an overestimation of k^{δ} (cf. Eq. (5.37)). These effects almost cancel each other out. Also, the thermodynamic parameters have only a marginal influence here. The change caused exclusively by the change in the applied microstructural parameters is only about 2 % for k^{δ}.

In contrast, the diffusion coefficient D^{δ} changes much more due to the higher value for τ_{LSCF} and the lower value for a determined from the reconstruction, both of which lead to an increase in D^{δ} (cf. Eq. (5.38)). Thus, the modified values for D^{δ} would be about 67 %

higher than the value calculated with the roughly estimated microstructural parameters in Ref. [2]. However, since the values used here for the thermodynamic factor are smaller than the values used in Ref. [2], the modified values for D^δ are only moderately higher (between 10 and 31 %, depending on the temperature), because lower values for γ_o lead to lower values for D^δ. Thus, the influence of the modified microstructural parameters and the modified thermodynamic parameters on D^δ act against each other.

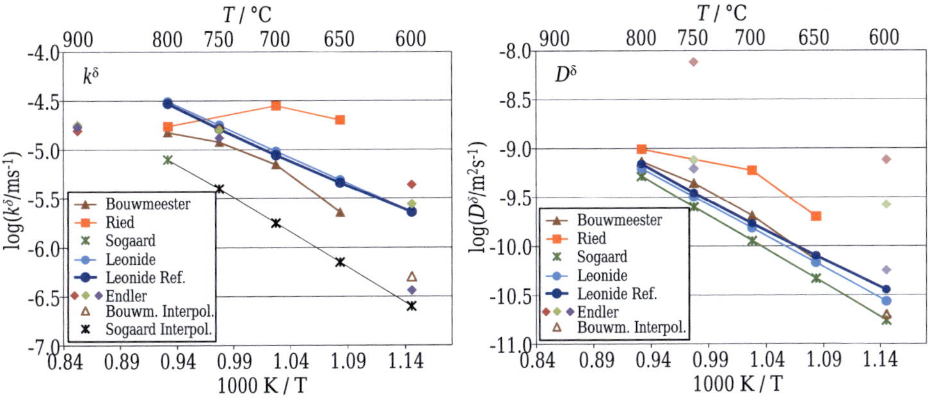

Figure 5.15.: Influence of the microstructural and thermodynamic parameters on the material parameters of Ref. [2]: (a) the surface parameter k^δ changes by only 3 %, while (b) the diffusion coefficient D^δ changes between 10 and 31 %. The values move towards the values calculated in Section 5.4.2 for $t = 1000\,\mathrm{h}$ (cf. Table 5.6).

As k^δ stays almost constant while D^δ increases, the penetration depth l_δ increases for the refined values. At $T = 600\,°\mathrm{C}$, the penetration depth increases from 1.16 µm to 1.34 µm, while at 800 °C, l_δ increases from 1.52 µm to 1.64 µm. However, it can be seen in Fig. 5.15 that the k^δ values are in good agreement with the values calculated in Section 5.4.2, especially at 750 °C where the value determined over 1000 h barely changes. For the D^δ values, it can be observed that the modified values are shifted in the direction of the values calculated over 1000 h. At 750 °C and 600 °C, the values match well to the values after 1000 h. Note, that the cells in [2] and [39] are operated under different conditions, which can influence the change in D^δ and k^δ over time. The cell in [2] was operated for approximately 350 h before characterisation.

The activation energies $E_a = 1.07\,\mathrm{eV}$ for k^δ and $E_a = 1.24\,\mathrm{eV}$ for D^δ changed only slightly due to the modification ($E_a = 1.05\,\mathrm{eV}$ for k^δ and $E_a = 1.26\,\mathrm{eV}$ for D^δ in Ref. [2]). An almost perfect match between the simulated and measured resistance can be gained by applying the refined k^δ and D^δ values together with $c_{O^{2-}}$ as in Table 5.3 for the simulations with ParCell3D using cathode 2_{AP} as computational domain. This is shown in Fig. 5.16, which underlines the validity of the parameters used to calculate k^δ and D^δ in this section.

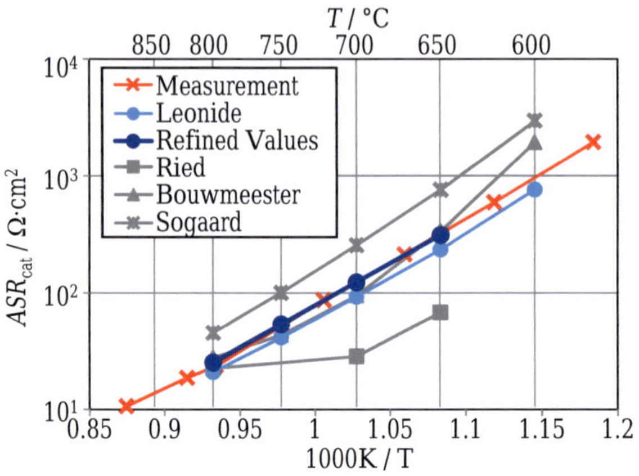

Figure 5.16.: The area specific resistance calculated with the modified k^δ and D^δ values determined in this section.

5.5. Accuracy Aspects of Calculating the ASR_{cat}

This section adresses accuracy considerations for simulating the ASR_{cat} with ParCell3D. First, the minimal size of an RVE with respect to the calculated area specific resistance is determined for the standard cathode from FZJ. Next, the influence of k^δ and D^δ dependent on the pO_2 inside the cathode is investigated. At the end of this section, a comparison is given between the Rüger model and ParCell3D.

5.5.1. Size of a RVE for Performance Simulations

The minimal size of an RVE for calculating the ASR_{cat} is approximated and discussed in analogy to Section 4.8.4. This means that the RVE is first calculated for a gradually increasing cubic volume (cf. Fig. 4.22a) and the results are plotted as a function of the number of particles per volume edge length L_{VE} (cf. Eq. (4.15)) and the total volume. Since the height of the electrochemically active volume might have an impact on the RVE size, this is calculated for two sets of material parameters k^δ and D^δ: one data set having a relatively shallow penetration depth of 1.44 µm (k^δ and D^δ from Ref. [2] at 750 °C) while the other has a relatively large l_δ of 7.32 µm (k^δ and D^δ from Ref. [39] at t_{start} at 750 °C). Additionally, nine volumes from different locations are analysed for $L_{VE} = 3.15, 6.3, 9.46, 12.6$ and 15.8. The standard deviation from the mean value are calculated and plotted in Fig. 5.17 (dotted lines). Note, that the same microstructures are used for both sets of k^δ and D^δ.

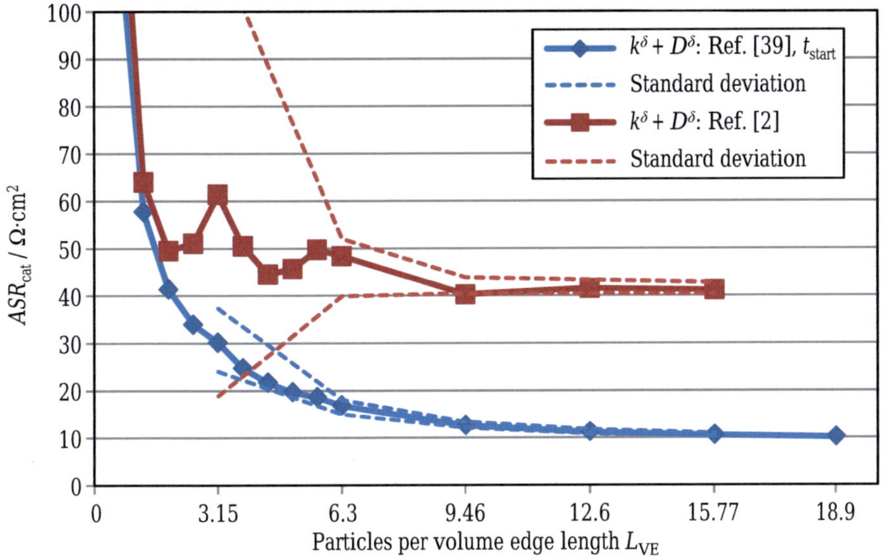

Figure 5.17.: Determination of a representative volume element (RVE) with respect to the ASR_{cat} for two different sets of k^δ and D^δ values (Refs. [2, 39] at 750 °C). The solid lines are calculated for one increasing volume, while the dotted lines indicate the standard deviation.

It can be seen that with increasing volume the ASR_{cat} decreases, whereas the curve calculated with the values from [39] (blue line) are much smoother compared to the (red) curve calculated with values from [2]. This behaviour is directly influenced by the penetration depth. For the blue curve, the penetration depth of 7.32 µm corresponds to L_{VE} of approximately 13.3. This means that up to 7.32 µm (or $L_{VE} \approx 13.3$), further increasing the volume increases the active surface area in all three directions and thereby lowers the ASR_{cat}. In contrast, the red curve has a penetration depth of only 1.44 µm ($L_{VE} \approx 2.6$) and thus the active area mainly increases in the lateral directions y and z for volumes with edge length > 1.44 µm, while the increase in x-direction (cathode height) does not play a significant role (the additional surface area in x-direction is not active). Hence, in the case of a shallow penetration depth, the increase of "active" volume is smaller than with greater penetration depth. Thus, an unfavourable location can have a much stronger influence on the calculated area specific resistance for the red curve, which could even lead to higher ASR_{cat} values with increasing volume (like e.g. at $L_{VE} \approx 3.15$). The standard deviations underline this finding, which can be seen by the larger standard deviations of the red curve (lower penetration depth) compared to the blue curve (greater penetration depth) in Fig. 5.17. Nevertheless, the standard deviations are low for both sets of k^δ and D^δ for $L_{VE} > 9.46$. It can also be seen in Fig. 5.17 that the ASR_{cat} is almost converged with $L_{VE} \geq 12.6$ for both sets of parameters and that both curves change only slightly afterwards (from $L_{VE} = 12.6$ to 15.8 the red curve changes by 0.8 % and the blue curve by 6 %). This indicates that for these volume sizes the results will be meaningful and representative for the whole structure. Nevertheless, the results show that for calculating the ASR_{cat} the RVE size can be anisotropic.

To further analyse the anisotropic nature of the RVE, two additional studies are performed. First, the base area of the analysed volume is increased for a fixed volume thickness of 8.75 µm, which corresponds to a thickness of about 12.6 particles. The results are plotted in Fig. 5.18a against the number of particles per edge length of the base area L_{BA} in analogy to L_{VE}. It can be seen that the ASR_{cat} is decreasing, similar to the study with the increasing *cubic* volume in Fig. 5.17. For the red curve (calculated with values from Ref. [2]) where the active area is much lower than the volume thickness, the percent change of the ASR_{cat} is similar to the case of increasing cubic volume (ASR_{cat} changes 4.3 % from $L_{BA} = 9.46$ to 15.8; in Fig. 5.17 it changed by 2.3 %). If using the values from Ref. [39], where the height of the active area is similar to the cathode thickness, the percent change of the ASR_{cat} is much smaller: the ASR_{cat} changes by 3.9 % from $L_{BA} = 9.46$ to 15.8 (in Fig. 5.17 it changed by 19 %). Between $L_{BA} = 12.6$ to 15.8, the red curve changes by only 0.2 %, while the blue line changes by 0.8 %. This emphasises that a base area of 12.6×12.6 particles ($7 \times 7 \mu m^2$) is sufficient, while even a base area as small as 9.5×9.5 particles will already give good results.

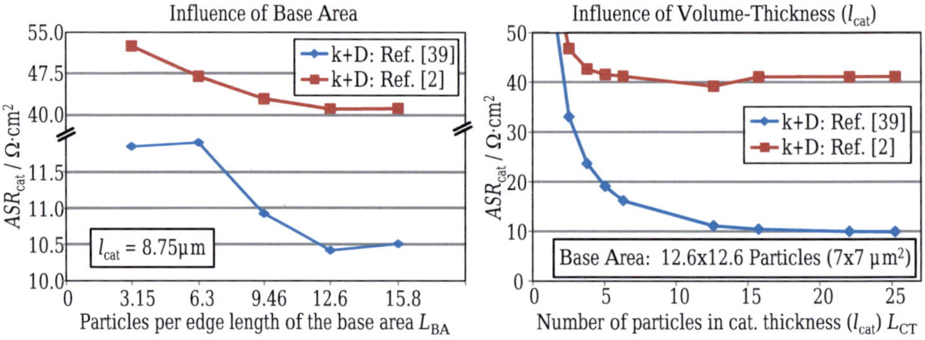

Figure 5.18.: Influence of the base area and height of the volume used to calculate the area specific cathode resistance for the determination of a RVE, with respect to the ASR_{cat}.

In the last study, the influence of the volume thickness was investigated by considering a fixed base area of 12.6×12.6 particles, while the volume thickness gradually increased. In this case it is expected that the ASR_{cat} will at first decrease with increasing thickness, as more surface area is provided for the incorporation of the oxygen. For cathodes thicker than the thickness of the electrochemically active area, a further increase will not lead to a further decrease of the ASR_{cat}, since the additional surface area is not active. Thus the ASR_{cat} will reach a minimum, from which the resistance will slightly increase again due to gas diffusion losses. This behaviour can be observed for both curves in Fig. 5.18b, where the results are plotted against the number of particles per cathode thickness L_{CT}. For the red curve calculated with values from [2], the ASR_{cat} converges quickly to a value of about 41 mΩcm^2 with increasing volume thickness. Already at a value of $L_{CT} = 5$ (which corresponds to a cathode thickness l_{cat} of 2.8 µm, the ASR_{cat} then only shows about 1 % difference to the value at $l_{cat} = 14$ µm ($L_{CT} = 25.2$). This is not surprising, since the penetration depth is only 1.44 µm. A similar behaviour is seen for the blue curve, although the ASR_{cat} converges much more slowly. From $L_{CT} = 5$ to 25.2, the ASR_{cat} changes by about 93 %. From $L_{CT} = 22.1$ ($l_{cat} = 12.25$ µm) to 25.2, the ASR_{cat} changes by only 0.7 %. This is again due to the much greater penetration depth compared to the other data set.

Overall it can be stated that the minimal volume needed to be representative for the calculation of the cathode resistance can show an anisotropic behaviour. Naturally, the volume has to be large enough to represent the overall microstructure (the microstructural parameters) accurately. But in contrast to the RVE size needed for the calculation of microstructural parameters in Section 4.8.4, a distinction must be made between the active and non-active volume if calculating the ASR_{cat}. How much of the cathode (starting from the electrolyte/cathode interface) is electrochemically active, depends on the ratio between D^{δ} and k^{δ} and the microstructure. It can be roughly approximated using Eq. (5.5), but it should be noted that the active volume is thicker than l_{δ}, which is only the point where the ion current dropped to 36.8 %. The volume above the active thickness shows almost no influence as long as the transport of oxygen in the pores and electrons within the material is not inhibited by the "inactive" microstructure (the gas diffusion losses at $pO_2 = 0.21$ bar are very small inside cathodes of less than 50 µm [37]). Thus it is not necessary to take the full cathode thickness into account, which would massively increase the computational effort. Hence, the height of the considered volume has to be as large as the active thickness of the cathode. To provide sufficient statistics, the size of the RVE's base area has to be adjusted depending on the active thickness, in order to consider a large enough (i.e. relevant) volume. It was found here that a base area of at least 12 to 15 particles should be considered for converging results for the ASR_{cat}. In conclusion, this means that the size of these RVE for calculating the ASR_{cat} does not only depend on the degree of structural homogenization, but also on the material parameters k^{δ} and D^{δ} used to calculate the cathode resistance.

5.5.2. Influence of pO_2 Dependent k^{δ} and D^{δ} Values

In the Rüger model (cf. Section 5.1.3) and the early version of ParCell3D, the k^{δ} and D^{δ} values used for the simulations are assumed to be constant within the cathode, as the oxygen partial pressure in the pores hardly changes and since the temperature is assumed to be constant. Nevertheless, to prove whether this simplification is justified, in the current version of ParCell3D the values for k^{δ} and D^{δ} are considered as dependent to the actual pO_2 (which is calculated spatially resolved inside the cathode), according to Eqs. (5.21) and (5.24). In these equations, the exponents α_k and α_D are the slope of k^{δ} and D^{δ}, respectively, with increasing pO_2. However, not much has been reported about the pO_2-dependent behaviour of k^δ and D^δ for LSCF. In Ref. [217], α_k is reported as 0.67, while the α_D can be calculated from the data published there as -0.332 for $La_{0.68}Sr_{0.3}Co_{0.2}Fe_{0.8}O_{3-\delta}$ at $T = 805\,°C$. In Ref. [196], α_D was found to be almost independent of the pO_2 ($\alpha_D = 0$) at $T = 800\,°C$ for $La_{0.6}Sr_{0.4}Co_{0.2}Fe_{0.8}O_{3-\delta}$, while the α_k was reported as 0.547. Since there is no consistent view in literature whether α_D is positive, negative or zero, this value is varied in the simulations. The k^δ and D^δ values are taken from Ref. [196] at $T = 800\,°C$. This point was chosen because the paper reports the pO_2 dependency of an LSCF composition almost identical to the one used here. Table 5.8 gives an overview on the performed simulations and the resulting ASR_{cat}.

Table 5.8.: The area specific cathode resistance ASR_{cat} and the penetration depth l_δ calculated at $T = 800\,°C$ and $p_{O_2} = 0.21$ bar with k^δ and D^δ values from Ref. [196].

k^δ / ms^{-1}	D^δ / m^{2}s^{-1}	α_k	α_D	ASR_{cat} / mΩcm^2	l_δ / µm
$7.91 \cdot 10^{-6}$	$5.17 \cdot 10^{-10}$	0	0	45.02275	2.33
$7.91 \cdot 10^{-6}$	$5.17 \cdot 10^{-10}$	0.547	-0.332	45.02533	2.33
$7.91 \cdot 10^{-6}$	$5.17 \cdot 10^{-10}$	0.547	0	44.98477	2.33
$7.91 \cdot 10^{-6}$	$5.17 \cdot 10^{-10}$	0.547	0.332	44.98480	2.33

As can be seen, the influence of α_k and α_D at $p_{O_2} = 0.21$ bar and $T = 800\,°C$ is negligible. All four simulations lead to an almost identical resistance, differing by less than 0.1 %. Also, the penetration depth is the same for all simulations. This can also be seen from the current profiles inside the cathode of the four simulations in Fig. 5.19. In this plot the dimensionless value of the current (the current is normalized by dividing it by its maximum value) with respect to the distance from the electrolyte (the cathode thickness) is shown. The ionic current takes its maximum value at the interface between MIEC and electrolyte (zero abscissa in the plot). The current decreases with the distance from the electrolyte and it can be observed that at a distance greater than 10 µm the current is below 1 % of its maximum value, while the penetration depth (distance where about 36 % of the current is reached) is given as 2.3 µm.

However, the differences in the current profiles are negligible. The course of the profiles calculated with $\alpha_D = 0$ and $\alpha_D = +0.332$ are identical, while the profile calculated with $\alpha_D = -0.332$ is slightly below the profile calculated with constant k^δ and D^δ values. Thus it can be concluded that for cathodes with a thickness of no more than 50 µm and a pO_2 of 0.21 bar, the pO_2 dependence of k^δ and D^δ can be neglected. Please note, the influence might strongly increase at lower pO_2, which can occur e.g. in a non-ideally contacted cell in a stack assembly due to oxygen starvation beneath the contact ribs.

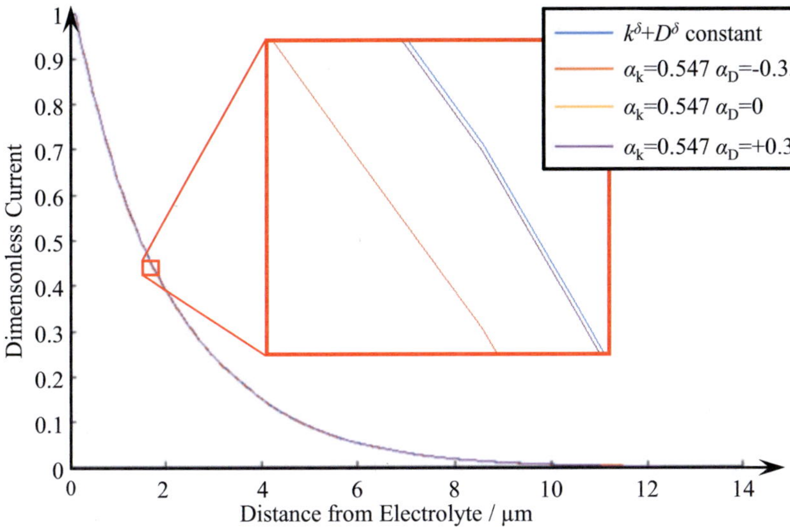

Figure 5.19.: Current profile of the simulations as performed with the parameters listed in Table 5.8 ($T = 800\,°C$, k^δ and D^δ values from Ref. [196]).

5.5.3. Comparison Between ParCell3D and the Rüger Model

As already mentioned, the principle difference between the Rüger model and ParCell3D is the representation of the microstructure as considered by the model. While in the Rüger model each particle or pore is represented by one cube, ParCell3D can cope with many more elements and is thus able to resolve microstructures much better (a particle can be approximated by many cubes or voxels). Hence, a comparison between the two models is a comparison between a simplified, coarse microstructure and a complex, high-resolution microstructure (very close to reality). For this reason, the Rüger model will also be referred to as *coarse model* in this section, while ParCell3D is referred to as *high-resolution model.*

A fair comparison of both models is not simple, as it depends on different factors. The biggest advantage of the high-resolution model (ParCell3D) over the coarse model (from Rüger) is its great flexibility, which allows it to consider any structure based on voxels as computational domain. Thus the comparison predominantly depends on how well the actual microstructure can be approximated by symmetrically aligned and equally sized cubes. In order to answer this question, the process of creating the microstructure for the coarse model should be understood. As discussed above, the main characteristics of MIEC microstructures are the surface area density a, the tortuosity τ_{LSCF} and the porosity ε. However, the cathode structure is approximated by cubes, and first the size (edge length) of the cubes l_v must be specified, which represents the size of the particles and pores in the model. Moreover,

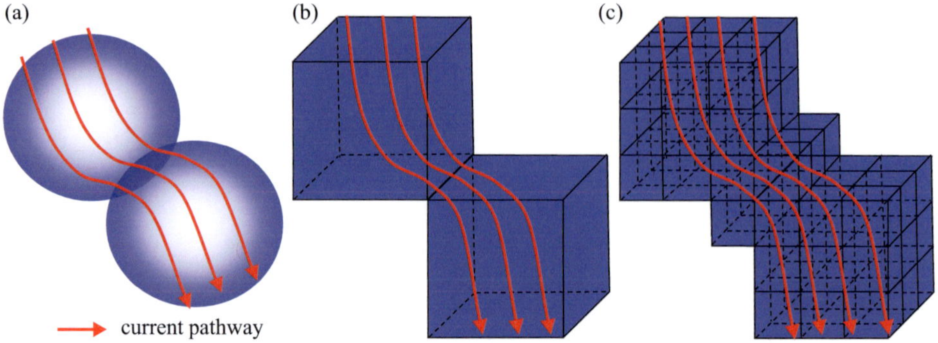

Figure 5.20.: (a) Scheme of two sintered, spherical particles and possible current pathways through the particles. (b) Approximating each particle by only one cube leads to two cubic particles with a shared edge. The current has to pass through the edge, if the pathway is to be maintained. (c) If the particles are resolved with a higher resolution, current pathways through the connecting cubes are possible.

the number of cubes in all three directions has to be set. It was shown in Ref. [26] that for the coarse model a base area of 7×7 cubes (or particles) delivers reliable results. The number of particles in the transport direction together with l_v define the considered cathode thickness. Also, the porosity ε and thus the material fraction $(1-\varepsilon)$ are specified and the corresponding percentage of cubes is randomly assigned as pores or MIEC material. This means that the surface area density a is determined only by l_v and ε. According to Ref. [3], the surface area of the coarse model a_{CM} can be approximated by

$$a_{CM} = \frac{5.812 \cdot (\varepsilon - \varepsilon^2)}{l_v}, \tag{5.41}$$

with an error $< 1\,\%$ (for $0.2 \leq \varepsilon \leq 0.5$). The tortuosities of both material and pores are determined only by the material fraction of the randomly assigned cubes in the coarse model. Since the actual material distribution can have an impact on τ_{LSCF} and a, the average value of 15 to 60 random material distributions is used. Note that current pathways over the edges and corners of the cubes are enabled in the Rüger model, as illustrated in Fig. 5.20). From a mathematical point of view, this does not make sense, since neither edges nor corners have a contact area where oxygen ions or electrons could pass. From the viewpoint of experience, allowing such pathways (which the COMSOL software automatically does) gives an intuitive approach for reproducing the microstuctural characteristics by such a coarse approximation, where one cube represents one particle. Enabling pathways between the particles, but avoiding the irregular pathways through edges and corners is only made possible by a dramatic increase in the number of cubes used for the simulations, which allows the structure to be resolved more accurately (cf. Fig. 5.20c).

Table 5.9.: Comparison between the microstructural parameters from cathode 2_{AP} (used in ParCell3D) and from the microstructure created for the simulations with the Rüger model. The ASR_{cat} is calculated with k^{δ} and D^{δ} values from Ref. [2].

	ParCell3D	Rüger Model
Av. Particle Size d_{LSCF} / nm	552	552
Av. Pore Size d_{pore} / nm	449	552
Porosity Fraction / %	44.6	44.6
Tortuosity LSCF τ_{LSCF}	1.68	1.60
Tortuosity Pore τ_{pore}	2.04	2.06
Surface Area Density a / µm^{-1}	2.88	2.60
ASR_{cat} at 600 °C / mΩcm^2	757	801
ASR_{cat} at 800 °C / mΩcm^2	20.7	21.3

However, with the given conditions of the Rüger model, it is possible to mimic the microstructural characteristics of MIEC cathodes with good agreement to the high-resolution model ParCell3D. The cathodes from FZJ are prime examples, as their microstructure is very homogeneous, the (average) particle and pore sizes are similar and they have relatively narrow and symmetrical phase size distributions (cf. Section 4.9). The microstructural characteristics of the coarse model's geometry (average values of 30 realizations with $7 \times 7 \times 26$ cubes), aiming to approximate cathode type 2_{AP}, are listed in Table 5.9. The values from the reconstruction of cathode type 2_{AP} are also listed for reasons of comparison.

The model geometry of the Rüger model was created by taking the average LSCF particle size to be the cube edge length ($d_{LSCF} = l_v$) and the porosity fraction ε as obtained from the reconstruction of cathode 2_{AP}. This determined the surface area density and the tortuosity of both phases, which show only a small difference to the reconstruction values. Hence, the ASR_{cat} differs by only 6 % at 600 °C using the k^{δ} and D^{δ} values from Ref. [2]. It is worth noting that the differences in the ASR_{cat} between the Rüger model and ParCell3D also depend on the applied k^{δ} and D^{δ} values and the ratio between these two parameters. This is because the surface area is linked to k^{δ}, while the tortuosity of the material is linked to D^{δ} (cf. Eqs. (5.37) and (5.38)). The structure with the smaller surface area density (in this case the coarse structure) has a larger penetration depth l_{δ} and thus a thicker active area (provided that τ_{LSCF}, ε, k^{δ} and D^{δ} are the same or at least similar). Hence, a change in D^{δ} will affect the coarse structure more than the highly-resolved structure. As D^{δ} from Ref. [2] does have a higher thermal activation energy of 1.26 eV compared to 1.05 eV for k^{δ}, the difference between the coarse and the high-resolution model decreases with increasing temperature. Broadly speaking, the quicker increase of D^{δ} compared to k^{δ} results in a faster drop of the ASR_{cat} for the coarse model with the thicker active area. Thus, the values for ASR_{cat} come closer to each other with increasing temperature. From $T = 600$ °C up to 800 °C, the difference between the two models decreases from 6 % to only 3 %.

To conclude, the coarse structure from the Rüger model gives similar results to those of the ParCell3D, obtained with a highly resolved structure, but only for homogeneous structures with narrow phase size distributions, where the average particle and pore sizes are close to each other. Beyond this, for more inhomogeneous structures, the differences between the two models drastically increase. Moreover, ParCell3D avoids irregular current pathways.

5.6. Influence of Microstructure on Performance

The influence of the microstructure on the ASR_{cat} will now be studied. For all calculations in this sections, the same k^{δ} and D^{δ} values (from Ref. [2] as in Table 5.1) are used. The simplest and fastest way to study the impact of the microstructure is by using the following equation

$$ASR_{\mathrm{cat}} \equiv R_{\mathrm{chem}} = \left(\frac{RT}{4F^2}\right)\sqrt{\frac{\tau_{\mathrm{LSCF}}\gamma_o^2}{(1-\varepsilon)a(c_{\mathrm{mc}})^2 D^{\delta} k^{\delta}}}, \tag{5.42}$$

which is Eq. (5.3) from the ALS model, written in terms of the chemical diffusion coefficient D^{δ} and the chemical surface exchange coefficient k^{δ}. The reason to use Eq. (5.42) instead of Eq. (5.3) is, that now the same values for k and D can be used in the 3D FEM model ParCell3D and in the ALS model. The detailed transformation between the two equations is shown in Appendix B.

The values used for γ_o and c_{mc} are calculated as described in Section 5.4 by using $C_1(T)$ and $C_2(T)$ as listed in Table 5.1. As also mentioned in Section 5.4, the thermodynamic factor is hard to determine exactly and can thus be a source of error when calculating the ASR_{cat} using Eq. (5.42). However, the main focus here is on the change of the ASR_{cat} (denoted as $\Delta ASR_{\mathrm{cat}}$) rather than on the exact value, and the percentage change is not affected by the applied value of γ_o.

First, the ASR_{cat} is calculated with Eq. (5.42) and the microstructural parameters as determined from cathode 2_{AP} ($a = 2.88\,\mu\mathrm{m}^{-1}$, $\varepsilon = 44.6\,\%$ and $\tau_{\mathrm{LSCF}} = 1.68$). Then each microstructural parameter is varied by increasing and decreasing its values by plus or minus 25 %. The change of the calculated cathode resistance due to this change in the microstructural parameters is shown in Table 5.10 alongside the applied values for a, ε and τ_{LSCF}. The parameters are changed by the same percentage in order to show the influence of each parameter on performance and to enable an easy comparison. At the end of the table, two calculations are also listed where all three parameters are varied at the same time in the direction which is (1) lowering or (2) enhancing the ASR_{cat}.

Table 5.10.: Influence of microstructural parameters on performance calculated with Eq. (5.42) from the ALS model using k^δ and D^δ from Ref. [2] (see Table 5.1). The reference value is calculated by using the microstructural parameters from cathode 2_{AP}.

a / µm^{-1}	Δa	ε / %	$\Delta\varepsilon$	τ	$\Delta\tau$	$ASR_{cat,750\,°C}$	ΔASR_{cat}
2.88	-	44.6	-	1.68	-	0.042 Ωcm^2	reference
3.60	+25 %	44.6	-	1.68	-	0.037 Ωcm^2	-10.5 %
2.16	-25 %	44.6	-	1.68	-	0.048 Ωcm^2	+15.5 %
2.88	-	55.7	+25 %	1.68	-	0.047 Ωcm^2	+11.9 %
2.88	-	33.4	-25 %	1.68	-	0.038 Ωcm^2	-8.7 %
2.88	-	44.6	-	2.10	+25 %	0.047 Ωcm^2	+11.8 %
2.88	-	44.6	-	1.26	-25 %	0.036 Ωcm^2	-13.4 %
3.60	+25 %	33.4	-25 %	1.26	-25 %	0.029 Ωcm^2	-29.3 %
2.16	-25 %	55.7	+25 %	1.68	+25 %	0.060 Ωcm^2	+44.5 %

As can be seen, the ASR_{cat} drops if the surface area is enlarged, or if either the tortuosity or porosity is reduced. Here, the tortuosity shows the biggest influence (ASR_{cat} drops by 13.4 %), while the ASR_{cat} is least affected by a change in porosity of the same percentage (ASR_{cat} drops by 10.5 %). If all the described parameter changes are made simultaneously, the ASR_{cat} would be reduced by 29.3 %.

On the other side, an increasing cathode resistance results from a decreased surface area (+15.5 %) or an increasing porosity (+11.9 %) and tortuosity (+11.8 %). Simultaneously changing all parameters in that direction would increase the ASR_{cat} by 44.5 %.

Despite the apparent simplicity of this study and the used model, this coarse estimation of the cathode resistance already highlights the influence of the microstructure on performance and shows the potential for lowering the ASR_{cat} by changing the microstructure. Nevertheless, in real, sintered microstructures, it is nearly impossible to change only one microstructural parameter while preserving the others. For example, the porosity is directly linked to the tortuosity. Lowering the porosity fraction naturally means an increase in the material fraction. This lowers the material phase tortuosity (and normally also lowers the surface area density, see Fig. 5.25), while increasing the tortuosity of the pore phase (leading to an increase of the gas diffusion losses, which is not considered in the ALS model).

To observe the influence of actual microstructures on performance, the 3D FEM model ParCell3D was used to simulate cathode performance by using the differently sintered microstructures of type 3 as computational domain. Thereby, the same material parameters as in the study above are used and the results are listed in Table 5.11.

The penetration depth l_δ increases with increasing sintering temperature due to the decreasing a and τ_{LSCF}. For reasons of comparison, Table 5.11 shows l_δ calculated with the ALS model (Eq. (5.5)) and with ParCell3D. Both models show the same tendency, but the values differ for the different cathodes, e.g. due to local inhomogeneities, thus demonstrating the differences between homogenized models (ALS) and spatially-resolved models (ParCell3D).

The calculated resistance is lowest for the cathode sintered at the lowest temperature (960 °C) and the ASR_{cat} increases with increasing sintering temperature. This is primarily due to decreasing surface area density as the sintering temperature increases. For cathode 3_{1200s}, the cathode resistance is over two times greater than for cathode 3_{960s}. Again, this shows the large influence of the microstructure on performance. However, for the measured resistances the differences between the cathodes are different, especially the cathode sintered at $T = 1200\,°C$, which showed an excessively higher polarization resistance compared to the other three cells [116].

Table 5.11.: The ASR_{cat} as simulated with ParCell3D ($ASR_{cat,sim}$) using the k^{δ} and D^{δ} values from Ref. [2] (see Table 5.1) and measured values ($ASR_{cat,meas}$) for the cathodes of type 3 (simulated/measured at 750 °C).

Cathode	a	ε	τ_{LSCF}	$l_{\delta,sim}$	$ASR_{cat,sim}$	$ASR_{cat,meas}$
3_{960s}	6.07 µm^{-1}	53.3 %	2.88	0.5 µm	27.48 mΩcm^2	56 mΩcm^2
3_{1030s}	4.86 µm^{-1}	53.3 %	2.43	0.5 µm	30.27 mΩcm^2	48 mΩcm^2
3_{1080s}	3.23 µm^{-1}	47.7 %	2.04	0.7 µm	37.68 mΩcm^2	133 mΩcm^2
3_{1200s}	0.75 µm^{-1}	24.9 %	1.23	4.9 µm	61.93 mΩcm^2	18 058 mΩcm^2

A fair comparison between the simulated and measured cathode resistance is unfortunately not possible, due to various reasons. Most important is the formation of secondary phases and interdiffusion layers at the interface between cathode and electrolyte [218–220]. The resistance caused by these secondary phases is not included in any microstructure model so far published, and are also difficult to separate from the measurement results. It should also be noted that all simulations performed here with different structures were done with the same values for k^{δ} and D^{δ} (determined in Ref. [2] from ASCs). Nonetheless, it is possible that the different sintering temperatures changed the material parameters (i.e. demixing or decomposition, etc.). This could be especially relevant to the structure sintered at $T = 1200\,°C$ with its excessive high polarization resistance. In contrast to the type 2 cathodes (which are from ASCs), the symmetrical type 3 cathodes contain secondary phases near the cathode/electrode interface, the contributions of which superimpose the Gerisher-impedance. Thus the method presented in Section 5.4 to determine k^{δ} and D^{δ} can not be used for these cathodes. The identification of the processes occurring in symmetrical cells, including the formation of secondary phases is ongoing work [218–220]. However, the discrepancy between simulations and measurements cannot be explained purely by differing microstructures.

Nevertheless, the results of this section clearly show the potential for performance increases through custom-tailoring the microstructure of a given material system. But the complete procedure of manufacture and then FIB tomography and microstructure model analysis is extremely time and labour intensive. Moreover, separating the different processes in the measured impedance spectra of such cells can be problematic. This shows the necessity of first studying the problem on a modelling level. There, purely microstructural influences on performance can be investigated - without any other effects like secondary phases, interdiffusion layers etc. Based on all these reasons, a model capable of generating synthetic microstructures with realistic microstructural characteristics would be highly desirable, as these structures can be used for a model-based optimization of the electrode structure. It is just such a model for the generation of synthetic microstructures that shall be presented in the next section.

5.7. Model for the Stochastic Generation of 3D Microstructures

This section presents a numerical tool capable of generating realistic yet synthetic porous microstructures for mixed conducting cathodes by imitating the manufacturing process. Before experimentally obtained microstructural data (e.g. from FIB tomography) were available, such models were used to calculate microstructural parameters or to obtain model geometries for adequate performance models, as discussed in Section 5.1. However, such a tool can also support the optimization of electrode microstructures, since the manufacture and reconstruction of electrodes are both extremely time- and labour-intensive. For example, parameter-studies in combination with adequate performance models (e.g. ParCell3D) can identify advantageous microstructural characteristics (cf. Section 5.8) and provide guidelines for custom-tailoring high-performance cathodes.

5.7.1. Short Overview on Existing Modelling Approaches

Various models were introduced in literature, since their results are very important for a model-based prediction of the electrode performance. Most models use a Monte Carlo approach to generate random packings or distributions of spherical particles. Some of these models also simulate the sintering process [187,189,221–223], which can be done in different ways. Ali *et al.* [221] and Cai *et al.* [189] consider the sintering process simply, enlarging perfect spheres to obtain a set degree of overlap between particles. Kenney *et al.* [222] vary the minimal distance between two contacting particles until the desired structural properties are obtained. Metcalfe *et al.* [223] add a sintering neck to particle intersection regions to capture the effect of sintering. Schneider *et al.* [187] use the discrete element method to numerically simulate the sintering process by mechanical interactions between the particles. Starting from an initial packing of non-overlapping spheres, the centres of the particles approach each other, thus imitating the densification which occurs during sintering.

However, the following subsections present a newly developed model for the generation of synthetic microstructures. In this model, first an initial configuration of particles is stochastically generated, which represents the cathode after screen-printing. Afterwards, the sintering process (cf. Appendix C) is simulated by rearranging particles and pores. Here, reducing the surface energy of the particles is considered to be the driving force. This allows a physically motivated manipulation of the surface area and the distribution of both phases. The approach is validated by comparing the microstructural parameters of the generated structures with those derived from microstructures reconstructed with FIB tomography. This will be explained in the following.

5.7.2. Applied Modelling Approach and Functional Principle

The presented model is entirely implemented in MATLAB [160]. It consists of two independent sub-models, mapping the manufacturing process of SOFC cathodes: The first sub-model creates an initial electrode microstructure, as expected after screen-printing. The slurry screen printed on top of the electrolyte (cf. Section 3.1) consists of ground up primary cathode material particles. Many popular cathode materials, such as the LSCF used in this study, are Perovskite crystals. Therefore, the ground-up particles of the initial microstructure are assumed to have a shape similar to the cells of a 3D Voronoi tessellation [224]. Consequently, the solid particles are created by employing a Voronoi tessellation. Hence, the particles of the initial structure have sharp edges and are just touching each other without any strong mechanical connection (the particles do not overlap; no sintering necks are present). This initial microstructure is represented by a voxel-based structure, such as the 3D image data obtained from FIB tomography.

To create the initial structure, the following input parameters are needed: First, the size of the considered volume has to be set (number of voxels in all three directions and the voxel size). Next, the desired porosity fraction ε and thus the material fraction $(1-\varepsilon)$ have to be defined. Finally, the size of the solid particles (and the pores) is determined by setting the number of particles which are considered inside the specified volume.

The working principle of this first sub-model is fairly simple: within the specified volume, a number of points (according to the number of particles which should be considered) are randomly distributed and afterwards a Voronoi tessellation is performed (cf. Fig. 5.21a). Cells are randomly assigned to the solid phase, until the desired material fraction is reached (cf. Fig. 5.21b).

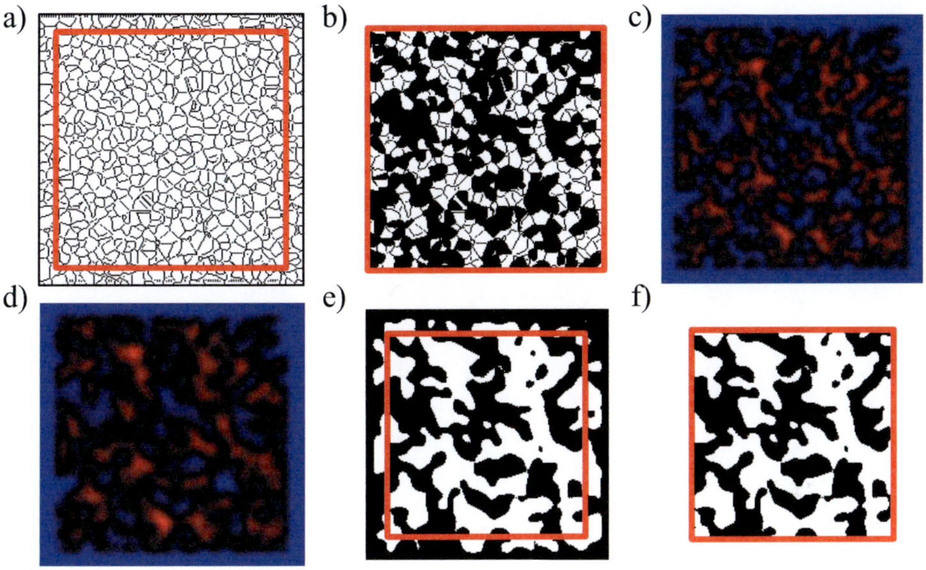

Figure 5.21.: Cross sections of all steps implemented in the model: (a) Voronoi tessellation (initial structure), (b) initial material distribution, cropping to avoid boundary effects (result of first sub-model). (c) A border of "pore voxels" (blue) allows for a proper calculation of the signed distance function in the volume of interest; (d) the signed distance function after numerical sintering; (e) after re-conversion in discrete voxel representation (result of second sub-model), where (f) the boundaries are cropped again to avoid boundary effects.

The second sub-model mimics the physical sintering process, in order to track its temporal influence on the cathode morphology. Although all numerically sintered structures in this thesis are created by the first sub-model using Voronoi tessellation, it is possible to "numerically sinter" all binary voxel-based structures with this model. This can also include packings of spheres or even segmented 3D image data obtained by FIB tomography.

One of the main causes of microstructural change during sintering is the reduction of surface energy, which results in a reduction of surface curvature (cf. Appendix C and Ref. [225]). To describe the evolution of a surface between two different domains, a level set approach [226] has been successfully applied in literature [138, 227, 228]. This leads to a smoothing of the surface of the particles and the formation and growth of sintering necks between neighbouring grains at the contact points (or contact areas). Thus, a good mechanical and chemical connection is established (cf. Appendix C).

In the presented model, the level set function $\varphi(x,t)$ is chosen as a signed distance function [226, 228] (cf. Fig. 5.23c and d). The evolution of the interface between pore and material can be modelled as proportional to the mean surface curvature $K_\varphi = \nabla \cdot \frac{\nabla \varphi}{|\nabla \varphi|}$, where $|\nabla \varphi| = 1$

holds for a signed distance function. The resulting differential equation that describes the sintering process is thus

$$\frac{\partial \varphi}{\partial t} + \nabla \cdot \nabla \varphi = 0, \tag{5.43}$$

where the temporal solution $\varphi(x,t)$ of this equation at a defined time t is the result of the second sub-model. This differential equation is solved using a built-in ODE solver from MATLAB. From the resulting sintered structure, a defined "boundary region" (typically 20 voxels at the outermost boundaries), is cut away to avoid any possible boundary effects. The actual sequence of the above described model steps is illustrated in Fig. 5.21.

Please note that t in Eq. (5.43) should not be considered as a specific time, but more as the *stage of sintering*. The larger t is, the higher the effect of the numerical sintering, which comprises both the effects of sintering time and temperature. For example, if the model imitates sintering for the same time at two different temperatures, the process with the lower temperature will have a smaller t value. Finding the optimal t for the imitated sintering stage is essential for the quality of the resulting structure and will be explained in the following.

5.7.3. Parametrization

The aim of this stochastic microstructure model is to create synthetic microstructures, which are as close as possible to real microstructures. To decide whether a synthetic structure is similar to a real structure, microstructural parameters of the created synthetic structures are calculated as explained in Section 4.7 and compared with the parameters obtained from 3D reconstructions. The microstructural parameters taken into account for the comparison are: porosity ε, median particle diameter d_{MIEC}, median pore diameter d_{pore}, surface area density a, tortuosity of the solid phase τ_{MIEC} and the tortuosity of the gas phase τ_{Pore}. The figure of merit is chosen to be the sum of the squared relative difference between the microstructural parameters of the model and those of the reconstruction.

In the first step, the model was parametrized to reproduce the state of the art cathode from Forschungszentrum Jülich. Thus, the goal was to create a synthetic structure with the same characteristics as cathode type 2_{AP}. For this aim, a parameter study had been conducted to identify the optimal set of parameters. The input parameters with the largest influence on the resulting synthetic structure are the number and shape of the particles in the initial structure (first sub-model) and the numerical sintering time t (second sub-model). The porosity fraction ε is given by the desired real structure, but it has to be noted that during the (numerical) sintering process, the porosity fraction can slightly decrease, because the material moves towards the centre of the structure and as the boundary regions of the structure are cut away. Thus, a slight structural densification results, as it is the case during actual sintering. This effect has to be considered during the creation of structures.

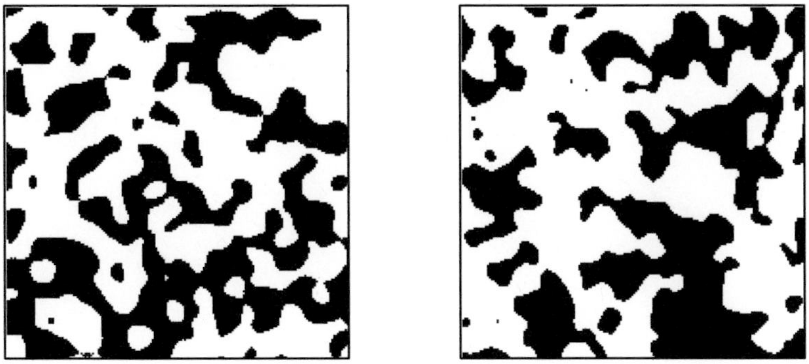

Figure 5.22.: Cross sections of modelled (left) and reconstructed (right) microstructures.

The ideal shape of the particles in the initial structure was identified by comparing different approaches with cells from a 3D Voronoi tessellation (using an Euclidean distance as metric). For example, initial structures based on spherical particles set surface area too high compared to reconstructed microstructures. The number of particles and pores can be roughly estimated by dividing the considered volume by the (average) volume of one particle or pore, which naturally depends on the desired (average) particle and pore size.

Table 5.12.: Comparison of microstructural parameters between reconstruction (type 2_{AP}) and model ($t = 5.5$, voxel size $= (35\,\text{nm})^3$; one Voronoi cell consists of $1.3 \cdot 10^4$ voxel).

	Porosity Fraction	Particle Size	Pore Size	Surface Area	Tort. MIEC	Tort. Pore
Reconstruction	44.6 %	0.55 µm	0.44 µm	2.88 μm^{-1}	1.68	2.04
Model	44.2 %	0.52 µm	0.46 µm	3.07 μm^{-1}	1.74	2.15
Rel. Difference	-0.9 %	-3.6 %	4.5 %	6.2 %	3.3 %	5.1 %

The most important input parameter is the numerical sintering time t, which can also be identified by a parameter sweep. Each imitated sintering stage is represented by a specific value of t. Thus, structures at different times t are quantified and the results are compared with the results from reconstructing cathode type 2_{AP}. The structure which has the greatest similarity to type 2_{AP} determines the sintering time t. Table 5.12 shows the quantitative comparison between model and reconstruction. Cross-sections of a reconstructed and a real microstructure are shown in Fig. 5.22, while in Fig. 5.23 a 3D illustration of both structures is given.

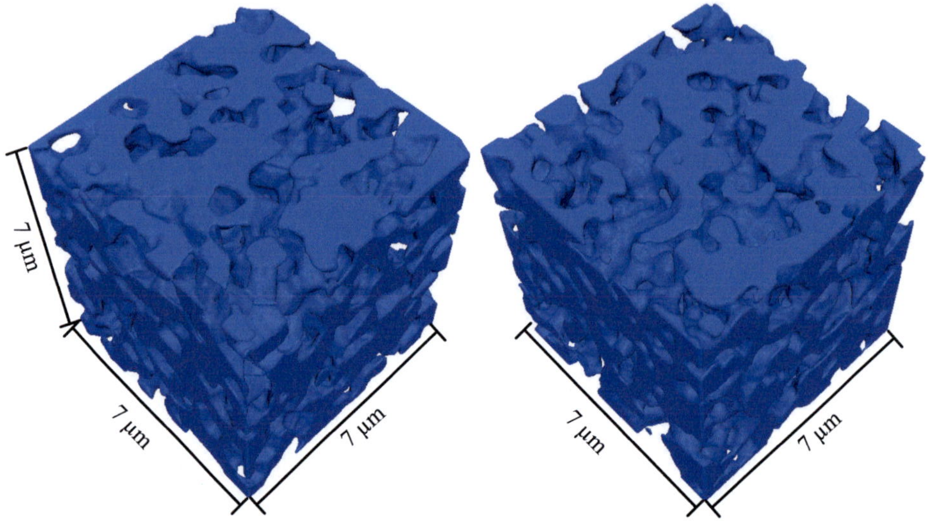

Figure 5.23.: 3D representation of modelled (left) and reconstructed (right) microstructure.

5.8. Optimization of the Cathode Microstructure

In this final section, strategies for optimizing cathode microstructure and lowering the ASR_{cat} are presented and discussed. Similarly to Section 5.6, cathode performance is studied alongside how it is influenced by microstructural characteristics (surface area density, porosity fraction and tortuosities of pore and material). Since a specific modification of microstructural characteristics in realistic microstructures is now possible, using the stochastic model presented in the previous section, the targeted microstructure optimization is attainable. Simulations with ParCell3D (cf. Section 5.2) will be performed in this section, all based on synthetic microstructures with desired microstructural modifications. For all simulations in this section, k^{δ} and D^{δ} values are used as published in Ref. [39] for $T = 750\,^{\circ}\mathrm{C}$ at the beginning of the measurements (not degraded; $\log(k^{\delta}[\mathrm{m/s}]) = -4.79$, $\log(D^{\delta}[\mathrm{m^2/s}]) = -8.12$).

Before optimizing the microstructure by selectively modifying it, the validity of the generated microstructures is tested. Simulation results performed with the reconstructed microstructure of type 2_{AP} are compared to those generated aiming to approximate cathode type 2_{AP}. The volumes for both structures considered for the simulations are $7 \times 7 \times 12.25\,\mu\mathrm{m}^3$ ($200 \times 200 \times 350$ voxels), which was found to be representative (see Section 5.5). The calculated ASR_{cat} for the reconstructed microstructure is $9.97\,\mathrm{m}\Omega\mathrm{cm}^2$, while for the generated cathode $10.08\,\mathrm{m}\Omega\mathrm{cm}^2$. Thus, the results of both structures differ by only about 1 %. It can therefore be concluded that the generated synthetic microstructure represents the reconstructed (real) microstructure well.

The first study of this section focuses on the surface area density. Next to the structure which approximates cathode type 2_{AP} with an average particle size of 552 nm, two more are considered: one has particles five times larger and the other has particles five times smaller. Thus, the porosity and the tortuosity are the same for all three structures, while only the surface area density is different. The ASR_{cat} is calculated as a function of the particle size and the cathode thickness. From the calculations plotted in Fig. 5.24, the influence of the surface area density on the performance can be studied, and the optimal cathode thickness approximated relative to the average particle size. The blue line is calculated from the structure approximating cathode type 2_{AP}, while the red and green curves are calculated for the five times lower and five times larger average particle size structures, respectively.

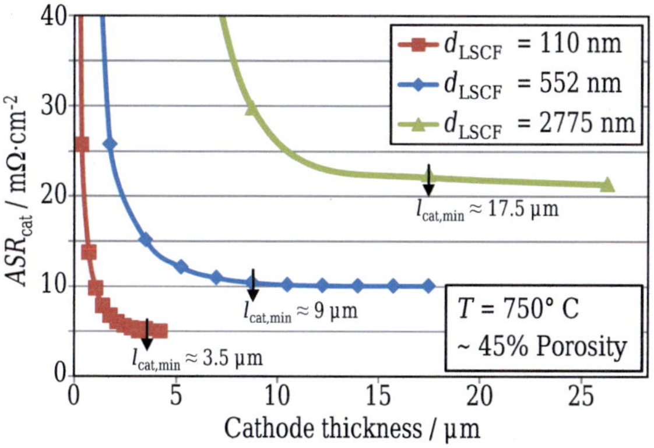

Figure 5.24.: The area specific cathode resistance as a function of the cathode thickness l_{cat}, calculated for different particle sizes d_{LSCF} with the k^{δ} and D^{δ} values from Ref. [39] at $t = t_{start}$ for $T = 750\,°C$.

First, Fig. 5.24 demonstrates that the ASR_{cat} of a MIEC cathode strongly depends on particle size, as the minimal value of the ASR_{cat} decreases from 21.2 mΩcm^2 for the 2760 nm particles structure to as low as around 5 mΩcm^2 for the 110 nm particles structure. This is caused by the increasing surface area density, which porous electrodes with smaller particle sizes provide for the oxygen surface exchange reaction.

Second, Fig. 5.24 shows that for a specific particle size, a lower limit is needed for the cathode thickness. This minimum cathode thickness $l_{cat,min}$ is for cathode type 2_{AP} with 44.6 % porosity (approximately 9 µm here). This electrochemically active volume of the MIEC cathode is adjacent to the cathode/electrolyte interface. From the studies in Section 5.5.1 it is known, that $l_{cat,min}$ also depends on the material parameters k^{δ} and D^{δ}. The larger the ratio of D^{δ}/k^{δ}, the larger $l_{cat,min}$. Moreover, it also depends on microstructural characteristics, especially on the surface area density. This can also be seen in Fig. 5.24, where the necessary cathode thickness is indicated as the minimum thickness $l_{cat,min}$. If calculating the penetration depth l_δ from Eq. (5.5), l_δ increases from 3.2 µm for the smallest particle structure, to 7.2 µm for the 552 nm particle structure and up to 16.0 µm for the largest particle structure. These values correspond well with the approximated minimal cathode thickness from the simulation results using ParCell3D shown in Fig. 5.24.

It should be mentioned that determination of the optimal cathode thickness can be difficult for cathodes built of particles smaller than about 100 nm, because gas diffusion losses from the added increment increase, resulting in a larger ASR_{cat}. However, and as already shown experimentally in [37], the gas diffusion polarization loss is negligible for cathodes with particles larger than 300 nm at current densities below $2\,\mathrm{Acm}^{-2}$ as is the case in these simulations. Due to the nonlinear current–voltage behaviour of the gas diffusion process, a radical increase of the polarization losses at significantly higher current densities is expected. However, it can be seen from this study that the ASR_{cat} approximately halves when the average particle size is reduced by a factor of about five. Naturally, in real cathodes the minimal realizable particle size that can be realized exists, which is predominantly determined by the material, the manufacturing process and the maximum operating temperature. It naturally occurs because small particles grow with increasing temperature.

In the next study, the influences of porosity and tortuosity are investigated. Both parameters are clearly linked to each other, since an increasing phase fraction typically goes hand in hand with a decreasing tortuosity of that phase. However, for this study microstructures were created with different porosity fractions between 5 % and 95 % (in 5 % steps). Thereby the stochastic microstructure model is again parametrized as for cathode 2_{AP}, which imitates a sintering process of about 3 h at 1080 °C. The microstructures are quantified as described in Section 4.7. The calculated surface area density is plotted in Fig. 5.25 with the tortuosities of LSCF and pores as a function of the porosity.

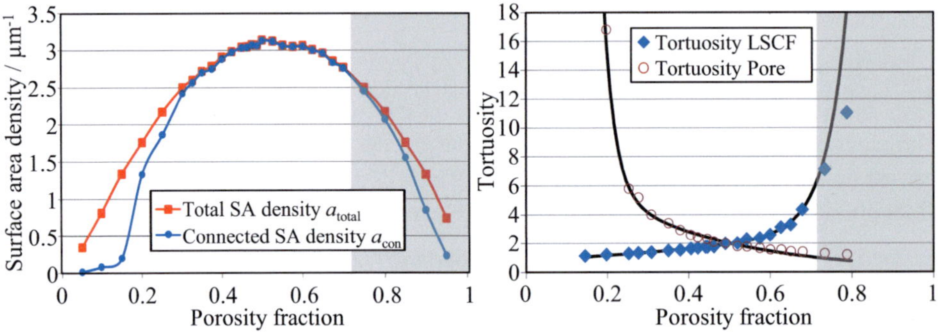

Figure 5.25.: Surface area density (left) and tortuosity (right) calculated for microstrctures with different porosities. The highest surface area density is reached at porosity of 50 % and the tortuosities of LSCF and pores are going in opposite direction.

First of all, it can be seen that the surface area density is largest for a porosity of around 50 % and decreases strongly for porosities below 30 % and above 75 %. Moreover, in this porosity range of 30 % to 75 %, almost all surface area is at open porosity, which means that pores are connected to the gas channel and, hence, almost all surface area is *connected* (see Section 4.7.1). Below 30 % the formation of isolated pores begins, because with decreasing porosity fraction some pores lose connection. Especially below 20 % of porosity, where the percolation threshold is reached, resulting in a clear drop of connected surface area. It should be noted that the curve for the open porosity is especially unsymmetrical; in contrast to isolated pores, isolated LSCF cannot occur in cathodes with only one solid phase (the LSCF material cannot hover in the pore phase). Thus unconnected surface areas can only result from dead-end material parts. However, fabricating structures with a porosity of more than 70 % would not make sense, due to stability reasons. Therefore the region with a porosity fraction of over 0.7 is greyed out in Fig. 5.25.

If looking at the tortuosity, it can be observed that the tortuosity of LSCF naturally increases with increasing porosity, while the tortuosity of the pores decreases. At about 50 % porosity, the two tortuosities are about the same. In the porosity range between 15 and 80 %, the two tortuosities can be described by the equations

$$\tau_{\mathrm{LSCF}} = 1.042 \cdot \exp(0.8615 \cdot \varepsilon) + 7.941 \cdot 10^{-4} \cdot \exp(11.86 \cdot \varepsilon), \tag{5.44}$$

$$\tau_{\mathrm{pore}} = 3.628 \cdot 10^{2} \cdot \exp(-10.82 \cdot \varepsilon) + 2.942 \cdot \exp(-1.012 \cdot \varepsilon). \tag{5.45}$$

The different behaviour of the tortuosity over the porosity clearly illustrates the compromise in optimizing cathode microstructure. However, as the transport of oxygen inside the pore phase shows lower losses compared to the transport of the oxygen ions inside the LSCF material, the tortuosity of the solid-phase is normally more important.

To examine this in more detail, the ASR_{cat} of each structure is calculated using the ALS model and ParCell3D. The results obtained for $T = 750\,°\mathrm{C}$ are plotted in Fig. 5.26 using the k^{δ} and D^{δ} values published by Leonide (as in Table 5.1). The calculated resistances for both models show the expected behaviour: at first they show a decreasing area specific resistance with increasing porosity, while later the resistance increases again. However, the two models disagree slightly about where the minimal resistance is located, although the overall accordance between the two models is very good. This accordance is because the synthetic microstructures are very homogeneous, typically even more homogeneous than real cathodes. For the calculations with the ALS model, it is important to use the connected surface area density a_{con} and not the total surface area density a_{total} (at least for low porosities; see Fig. 5.25 and Section 4.7.3), because only these connected surfaces can contribute to the oxygen reduction reaction. However, the structure with about 35 % porosity shows the lowest cathode resistance, if calculated with ParCell3D. If calculated with Eq. (5.3) obtained from the ALS model, the resistances are lowest for structures with a porosity of about 25 %. In general, the discrepancy between both models increases for lower porosity fractions. This is mostly as the gas diffusion losses increase at lower porosities and, in contrast to ParCell3D, the gas diffusion losses are not considered in Eq. (5.3). Thus in this porosity range, the results from ParCell3D are more reliable and the ideal porosity fraction is assumed to be about 35 %.

In conclusion, the ideal structure for a low cathode resistance is one with 35 % of porosity. The particles should be as small as possible, but large enough that manufacturing and operation do not alter the intended microstructure. In the future, the stochastic microstructure model can be used in combination with the performance model ParCell3D e.g. to study the influence of manufacturing on microstructural characteristics and performance by changing the input parameters of the stochastic model (e.g. grain size, sintering temperature and time). Thus, the manufacturing process can be optimized by identifying the ideal manufacturing conditions.

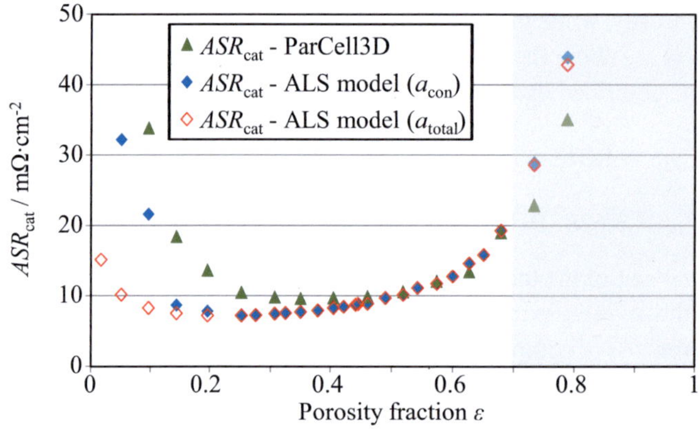

Figure 5.26.: The area specific resistance ASR_{cat} as a function of porosity, calculated with the ALS model and ParCell3D, respectively. The calculations are done for $T = 750\,^{\circ}\text{C}$ with the k^{δ} and D^{δ} values from Ref. [39] at $t = t_{\text{start}}$.

6. Summary

This section summarizes the results and the main findings of this work. Since not all details can be discussed here, references to the corresponding sections of this work are given as appropriate. This thesis dealt with the microstructural characterisation and simulation of porous, mixed ionic and electronic conducting (MIEC) cathodes for Sold Oxide Fuel Cells (SOFCs). Its main aim was to better understand the complex relationships between electrode microstructure, material composition, long-term stability and performance. This understanding would lay the foundation for microstructure optimization. Porous electrode microstructure is, next to material composition, mainly responsible for the performance of SOFC single cells. It is therefore essential to separate the material and microstructural influences on performance, to improve the understanding of internal electrode processes and their relationship to the microstructure.

Detailed knowledge of the 3D microstructure allows a correlation of the microstructure and electrode performance. To this aim, focused ion beam (FIB) tomography was adapted for reconstructing SOFC electrode microstructures. FIB tomography can collect a sequence of 2D micrographs, which are then reconstructed in 3D space and subsequently used for a detailed microstructure characterisation. An overview of the global developments and the current state of SOFC electrode reconstruction via FIB tomography was given (Section 4.1), enable to compare this thesis work to the context of other research groups worldwide. The presented work has led to many improvements in data acquisition and image processing, and has also been adopted by other groups. First, the epoxy resin infiltration of porous electrodes was optimized. This led to significant benefits for a fully automated reconstruction process and for image processing (Section 4.2). The conventional bulk milling technique (cf. Fig. 2.9) was improved by polishing two cross-sections perpendicular to each other, so that the sectioning process can start right from the edge of the sample (cf. Fig. 4.4). Combined with milling a trench either side of the analysed volume, it was shown that this method minimizes shadowing effects, problems with re-deposition and prevents brightness gradients between consecutive SEM images. An advanced electrode reconstruction approach using the information obtained from two detectors was introduced (Section 4.5). This is especially important when analysing structures that contain isolated pores and multi-phase electrodes, since each detector can be optimized for the identification of different phases.

Simultaneously to advances in sample preparation and data acquisition, analytical capabilities were established in image processing and geometrical analysis of the microstructures.

Robust phase identification is essential for an accurate microstructural analysis, thus two segmentation methods were discussed. The first method is the widely applied threshold segmentation, whereby appropriate methods for identifying the optimal threshold value were evaluated. Second, a self-developed segmentation algorithm was presented, belonging to the region growing segmentation methods. It allows automated greyscale segmentation, even for SEM images with low contrast between the phases. Although this algorithm was introduced for the segmentation of three phases, it can be easily applied to more or to fewer.

Accurate segmented data represent the actual electrode microstructure and allow a quantitative characterisation. Microstructural parameters were calculated, which enabled a structural comparison of different electrodes or revealed whether the porous electrode microstructure changes during operation. The techniques and methods for calculating the volume fractions, surface areas, tortuosities and phase size distributions of SOFC electrodes were carefully evaluated. They represent the consistent further development and application of existing methods.

Much of this thesis aimed to investigate and remove potential error sources in the reconstruction process, from sample preparation up to final parameter identification. Very little has been previously published. The efforts presented in this work covered the preparation process, the segmentation and the methods for parameter calculation, etc. Moreover, a detailed error analysis on volume representativeness, re-sampling and resolution of the 3D image data and their effects on the accuracy of the calculated parameters were presented and guidelines were given (Section 4.8). For example, for cathode type 2 (average particle size: 552 nm) it was found that a volume containing 15^3 particles (about 580 μm^3) provided already good statistics and can thus be defined as representative volume element (RVE). Moreover, a resolution of 10 to 15 voxels per particle-diameter was found to be the minimum. These findings can be used in future studies to specify the settings for an accurate reconstruction.

Considering all these findings, it was possible to reconstruct various cathode types with low resolutions (25 to 35 nm voxel size) and the largest volumes so far reported in literature (up to 32 049 μm^3). This allowed the detailed and accurate reconstruction, and subsequent quantification, of different structures.

It was thereby demonstrated that microstructure hardly changes during operation for more than 1000 h, whether operated at 600, 750 or 900 °C (type 2 cathodes in Section 4.9). Microstructural changes related to fabrication sintering temperature were also quantified (cathodes type 3). With increasing sintering temperature the porosity fraction, the surface area density and the tortuosity of the material decreased, since denser structures with larger particles resulted.

Adequate performance models are necessary for separating and quantifying material and microstructural influences on performance. An overview of published models was given in Section 5.1, with special focus on MIEC cathode models. Here the microstructure must be considered as accurately as possible. Accordingly, it is best to use the reconstructed microstructures for the simulations directly. A 3D FEM performance model called ParCell3D was presented, which uses the reconstructed microstructures as computational domain for calculating the area-specific cathode resistance ASR_{cat} as a performance index (Section 5.2). This model was developed in collaboration with T. Carraro (University of Heidelberg) and it is a further development of the model of Rüger (Section 5.1.3). Using this model, we were the first to do performance simulations using detailed 3D tomography data of the complex cathode microstructure. Calculating the cathode resistance of a given material system in dependency of its actual microstructure can isolate the influences on performance.

In ParCell3D, like in other MIEC cathode performance models, the material is represented by the chemical diffusion coefficient D^{δ} and the chemical surface exchange coefficient k^{δ}. The influence of these parameters on the calculated ASR_{cat} was investigated (Section 5.3), showing that the wide scatter of literature values of these parameters naturally lead to a large scatter in the calculated ASR_{cat}.

Therefore, a highly accurate method was presented for evaluating these material parameters from porous MIEC cathodes (Section 5.4). This contrasts against most published methods which use dense bulk samples. The evaluated method enabled to track the change of k^{δ} and D^{δ} during operation for the first time. Combined with the results from the microstructural quantification before and after operation, it could be shown that the degradation is almost exclusively caused by the material: at the operating temperature of $T = 600\,°C$ the significant degradation was caused by degradation of both k^{δ} and D^{δ}. At $T = 750\,°C$ a smaller degradation than at 600 °C was observed, mostly caused by a nonlinear decrease of D^{δ} by one order of magnitude, while k^{δ} barely changed. Only a small degradation could be observed at 900 °C, which could again be attributed to the oxygen ion diffusion in the bulk, described by a slightly increasing D^{δ}.

The material parameters k^{δ} and D^{δ} were also evaluated using measurement results from A. Leonide (Section 5.4.3), and showed good agreement with the previously calculated values. Moreover, it could be shown that when using these values to calculate the ASR_{cat} with ParCell3D based on the reconstructed cathodes as computational domain, the simulations matched near perfectly to the measurement results, indicating that the parameters used for the calculations were correct.

A detailed analysis of volume representativeness when calculating the ASR_{cat} was also performed. It showed that a RVE for the calculation of the ASR_{cat} can be anisotropic. Firstly, the volume has to be large enough to accurately represent the microstructural parameters. But moreover, the height of the considered volume must contain the full active thickness of the cathode. The volume above the active thickness shows almost no influence, as long as the transport of oxygen in the pores and electrons in the material are not inhibited. How much of the cathode is electrochemically active (starting from the electrolyte/cathode interface), depends on the k^{δ} and D^{δ} values and the microstructure. Depending on the active thickness, the considered base area could be smaller (in the case of great active thickness) or larger (in the case of low active thickness; see Section 5.5.1).

The influences of microstructure on cathode performance were also investigated. It was shown that from the best to the worst microstructure, fabricated by changing the sintering temperature, the calculated ASR_{cat} increased by about a factor of two. However, by comparing the simulation results with measurement results, it became obvious that when one wants to study only the influence of microstructure on cathode performance and exclude other influences (e.g. disregarding secondary phases at the cathode/electrolyte interface), it is necessary to first study this at the modelling level. To this aim, a tool was developed which can create realistic yet synthetic microstructures (Section 5.7). This stochastic microstructure model mimics the sintering process and was shown to create synthetic microstructures with microstructural characteristics near identical to the reconstructed microstructures. This was done to validate and parameterize this model.

Based on this stochastic model, realistic microstructures with differing microstructural characteristics were created. These structures were used to calculate the area specific resistance and identify microstructural characteristics advantageous for high performance. It was shown that the ASR_{cat} approximately halves when the average particle size is reduced by a factor of about five. By creating microstructures with different porosity fractions, it could be shown that for good pore-phase connection (i.e. negligible amount of isolated pores and thus also of isolated surface area) minimum porosity is 20 %. The surface area density was largest between about 40 and 60 %. The optimal porosity was identified at about 35 %.

Many practical applications are feasible for the presented methods. For example, they could optimize the fabrication process by pinpointing the influence of different input fabrication parameters on the microstructure, or to control the microstructure during large-scale production with an established fabrication procedure. Also, quantifying microstructural changes in the electrodes due to different operation conditions is possible and will help to better understanding the relationships between electrode microstructure, material, long-term stability and performance.

By achieving the goals set out in this work, the model-based optimization of microstructure is within reach.

Appendix

A. Reconstruction of Ni/YSZ Anodes

The basic principles and methods developed and used in this thesis are independent from the type of electrode, which is reconstructed and analysed. However, there are some particularities when reconstructing electrodes with three phase (e.g. Ni/YSZ anodes) compared to electrodes consisting of two phases (e.g. LSCF cathodes). These particularities will be mentioned in the following, if not already mentioned before. In addition, results of the reconstruction of Ni/YSZ anodes will be presented and discussed in the following.

A.1. Anode Samples and Aim of Study

This subsection provides information about the reconstructed anode samples and introduces the aim of the studies. Optimizing Ni/YSZ anodes is even more complex than optimizing the microstructure of LSCF cathodes, as they consist of three phases (nickel, YSZ and pore). All manufacturers have their own concepts and tries to optimize the anode layers differently. Therefore cells of different manufacturers are characterised, as will be explained below. All anodes analysed within this thesis are from ASCs consisting of a functional layer (AFL) and a substrate. Both layers are designed differently. Four different types of ASCs from different manufacturers are analysed in the following. The four cells are denoted as type A, B, C and D, respectively. The most obvious differences are the thickness of the two layers and the material fractions of the different phases inside the layers. All anodes are Ni/YSZ composite-anodes, but it has to be noted that type A, B and C contain 8YSZ (in the substrate and the AFL), while the substrate of type D contains 3YSZ (cf. Section 2.3). As the layers were produced from different powders, also the grain- and pore sizes of the three phases are supposedly different. The main characteristics of the samples are listed in Table A.1.

The first aim of this study is to analyse and discuss the differences between the AFL and the substrate. As the two layers have a different purpose, their microstructure should be designed differently. Hence it is of importance to analyse and optimize both layers separately. The second aim is to quantify and compare the different anode types and discuss the differences between them. The optimization strategies of the manufacturers shall become obvious and an evaluation on how good the microstructures actually are should be enabled.

Table A.1.: Design of the four anode-supported cells reconstructed and quantified in this work, with their layer thicknesses and their treatment before reconstruction via FIB tomography.

Sample	l_{AFL}	l_{sub}	Treatment before FIB Tomography
Type A	12 µm	1500 µm	~300 h, $T_{op} = [650 - 850°C]$
Type B	7 µm	500 µm	only reduction at 800 °C
Type C	7 µm	1500 µm	~400 h at 750 °C; 500 mA/cm^2 load
Type D	16 µm	500 µm	~140 h at 750 °C; OCV

A.2. Particularities of Parameter Calculation with Three Phases

In general the same methods used to calculate the microstructural parameters of LSCF cathodes can also be used for Ni/YSZ anodes, with some few modifications. Furthermore, for electrodes containing three phases, the triple-phase boundary length is of importance. This aspects will shortly be discussed here.

Phase Connectivity or Percolation

YSZ-voxels are considered connected if percolating with the electrolyte, Ni-voxels if percolating with the current collector and pore-voxels if percolating with the gas channel.

Surface Area

From the surface areas of the individual phases, the *interfacial* surface area (surface area between two phases) can be calculated by solving the following system of linear equations (for Ni/YSZ anodes: $n = 3$):

$$\begin{aligned} A_{\text{phase1}} &= A_{\text{phase1/phase1}} + A_{\text{phase1/phase2}} + \dots + A_{\text{phase1/phase n}} \\ A_{\text{phase2}} &= A_{\text{phase2/phase1}} + A_{\text{phase2/phase2}} + \dots + A_{\text{phase2/phase n}} \\ &\vdots \\ A_{\text{phase n}} &= A_{\text{phase n/phase1}} + A_{\text{phase n/phase2}} + \dots + A_{\text{phase n/phase n}} \end{aligned} \tag{6.1}$$

Triple-Phase Boundary (TPB) Density

The triple-phase boundary length is an important parameter for characterising and comparing SOFC electrodes consisting of an electronic, ionic and porous phase. As discussed in Section 2.4 the oxidation reaction of Ni/YSZ anodes occurs on or near the TPB where Ni, YSZ and the pore phase meet. Thus, the TPB in 3D images of Ni/YSZ anodes are voxel-edges that simultaneously touch voxels of the three different phases Ni, YSZ and pore. The most intuitive way of calculating the total TPB length L_{TPB} is to calculate it directly from the voxel mesh by a simple edge summation:

$$L_{\mathrm{TPB}} = N_{\mathrm{TPB,ve}} \cdot l_{\mathrm{v}} \tag{6.2}$$

where $N_{\mathrm{TPB,ve}}$ is the number of voxel-edges identified as TPBs and l_{v} is the edge length of the cubic voxels. However, this method will lead to a strong overestimation of the TPB length due to the steplike pattern of voxel structures, as the "real" TPB is much smoother. In a simple example where the TPB is a circle (assuming two overlapping spherical particles - one is nickel and the other YSZ - livitating in a pore: the TPB is a circle), the TPB length would be overestimated by more than 50 %, depending on the resolution and the degree of overlapping of the spheres, as for example shown in [126]. In literature different methods for calculating L_{TPB} were presented. A short overview is provided by Iwai *et al.* in [97]. A comparison of different methods and a discussion of the advantages and disadvantages can also be found in [139].

A simple method is to calculate the TPB length as described above by edge summation and correcting the value by a certain factor (e.g. -20 % in [190]). Shikazono *et al.* [126] introduced two alternative methods for calculating the TPB length in addition to the calculation by the TPB edge summation. The *midpoint method* defines the connection length of the midpoints of the TPB edge segments. An even smoother TPB is obtained by the *centroid method* which gives the total distance between the centroids of the triangles defined by the neighboring midpoints of the edge segments [126] as illustrated in Fig. A.1. Unless otherwise specified, the centroid method is applied in this work.

Table A.2.: Difference between the TPB density if calculated by a simple edge summation or by the Centroid Method [126] at the example of the total TPB of anode type A. As can be seen the calculation based on the voxel mesh is almost 50 % larger than with the centroid method.

Method	Anode Functional Layer (AFL)	Substrate
Voxel-mesh (VM)	$3.94\,\mu m^{-2}$	$2.71\,\mu m^{-2}$
Centroid Method (CM)	$2.56\,\mu m^{-2}$	$1.76\,\mu m^{-2}$
Factor between VM and CM	1.54	1.54

Analogous to the surface area, the TPB length is expressed in terms of a TPB density in order to enable a comparison of structures with different volumes. Hence, the triple-phase boundary density l_{TPB} is the calculated TPB length (L_{TPB}) divided by the reconstruction volume. Thereby the volume which has to be used for this normalization is the same as defined in Eq. (4.7).

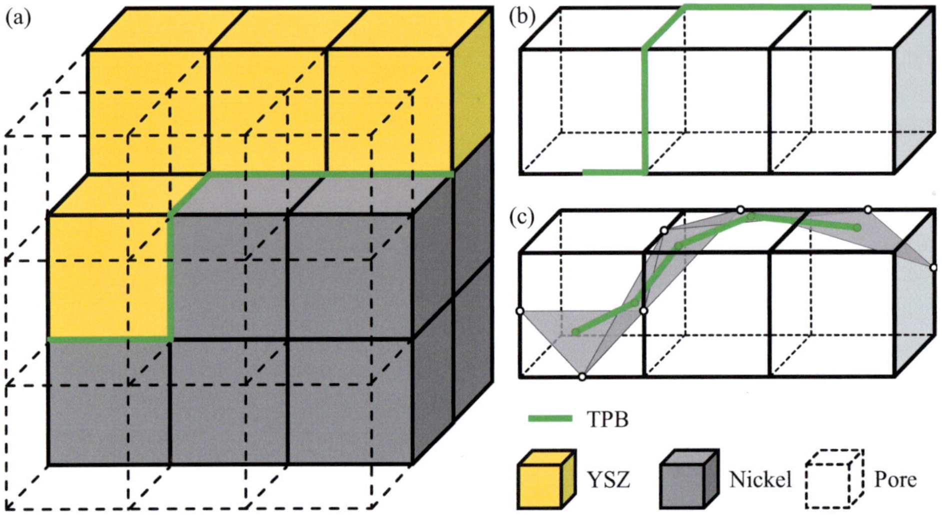

Figure A.1.: (a) An exemplary material distribution of a Ni/YSZ anode and the resulting TPB length if (b) calculated directly from the voxel mesh by summing the TPB edges and (c) by using the *centroid method* (adapted from [126]).

A comparison of the TPB density calculated by a simple edge summation and with the centroid method is given in Table A.2. There, the total TPB for the anode functional layer and substrate of anode type A is shown. The edge summation method leads to much higher values (~1.54 times higher) compared to the centroid method, where the TPB is much smoother and thus more realistic. The factor between the values obtained from different methods depends on the shape of the particles, or more precisely on the connection between the different phases at the TPB. The huge difference between these values clearly shows the necessity of applying appropriate methods. Moreover it aggravates a comparison of literature data.

Using the results of the connectivity analysis where all voxels were labelled as *connected*, *isolated* or *unknown*, the TPBs can also be categorized to one of these categories. Thereby, a TPB edge is only considered *connected*, if the voxels of all three phases (which are touching this TPB-edge) are labelled *connected* (cf. Section 4.7.1). If at least one of these phases is isolated, these TPB-edge is labelled *isolated*. All other TPB-edges are labelled as *unknown* (no phase is considered *isolated* and at least one phase is considered *unknown*).

A.3. Size of Representative Volume Elements for Ni/YSZ Anodes

As for LSCF cathodes, the minimum RVE size will be evaluated for Ni/YSZ anodes. In Ref. [229] it is mentioned, that an error of <2 % was found for electrodes with three-phases for volumes above 500 µm^3. However, it is not mentioned there how large the particles are. The RVE size for our structures will be investigated in the following, similar to the studies presented in Section 4.8.4 for LSCF cathodes.

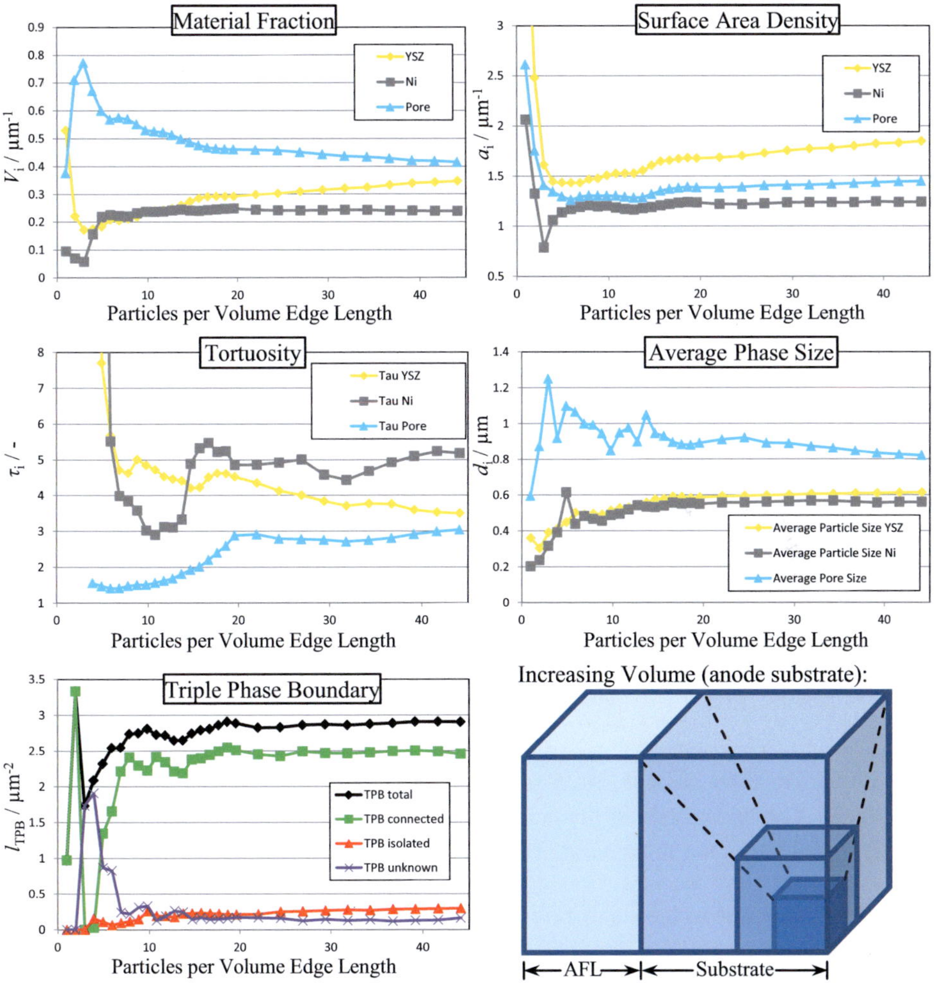

Figure A.2.: Calculated microstructural parameters for an increasing volume: material fractions, surface area densities, tortuosities, average phase sizes and triple-phase boundary density of the substrate of anode type C.

The RVE size is analysed for the anode functional layer and the anode substrate separately: first, for the substrate of anode type C (as this constitutes the largest data set of all anodes) and afterwords for the AFL of type D. As for the cathodes, the microstructural parameters are calculated for an increasing volume (see Fig. A.2).

If looking at the material fractions of the individual phases, it can be seen that for small volumes ($L_{VE} < 5$) the shares of the individual phases quickly changes. While the amount of nickel stays almost constant for larger volumes ($L_{VE} > 6$), the shares of YSZ and pores are slightly but constantly changing until the largest analysed volume of ~7800 µm^3 ($L_{VE} = 42$). This means that even a volume of 8748 µm^3 might be too small to be representative for the whole anode substrate. This would support the findings of Laurencin *et al.* [85], who stated that for their substrate a volume as large as about 42 800 µm has to be considered for an accurate determination of the porosity fraction using X-ray tomography. However, although their study was performed with another anode substrate and the study was done via X-ray tomography instead of FIB tomography, it clearly shows that for electrodes with three phases, the RVE size is much larger compared to an electrode with only two phases (e.g. LSCF cathodes).

Another explanation for the constant change of the pore and YSZ fraction is that the YSZ content constantly decreases from the AFL towards the gas channel, while at the same time the porosity fraction increases. The analysed volume increases mainly from the gas channel towards the AFL in this study. As the YSZ fraction increases for larger analysed volumes, it means that there is a higher YSZ fraction towards the AFL. Interestingly, the nickel fraction is almost constant.

A similar picture emerges when looking at the surface area densities. For volumes smaller than 5^3 particles, the surface area densities change fast with increasing volume, but only very slightly afterwards. The strongest increase in surface area can be seen for the YSZ, which is due to the increasing YSZ fraction in this region. However, for volumes bigger than about 15^3 particles (about 885 µm^3), the surface area densities (at least of nickel and pores) stay almost constant.

The tortuosity values show the highest dependency on the size of the analysed volume. The tortuosity of the pores changes fast before a volume of 20^3 particles, but only very slightly afterwards. This indicates that the pores are well connected and the paths are relatively homogeneous across the structure. Another behaviour can be seen for the tortuosities of the two solid-phases, as both values barely converge. Especially the course of Ni tortuosity is unstable. This is due to the small content of nickel (and to less extent of YSZ), which is near or below the percolation threshold (at ~30 % in case of three phases, see e.g. Ref. [26]). For example, between L_{VE} of 13 to 14, the tortuosity of nickel strongly increases due to a current constriction in this area. However, for very large volumes (7000 to 8748 µm^3) the tortuosity values are only barely changing and might already converged.

For the anode substrate the average phase size is one of the most stable parameters regarding the analysed volume. For volumes larger than 15^3 particles, the average phase sizes stay almost constant. Only the average pore size slightly decreases.

The parameter the least sensitive to the analysed volume in this study is the triple-phase boundary density l_{TPB}. It strongly changes before L_{VE} of 10 and less pronounced afterwards. For volumes larger than 20 particles per volume edge length (about 1700 μm^3), the value stays constant. This holds not only for the total amount of TPBs, but also for the TPB shares identified as *percolating*, *unknown* or *isolated*.

It can be concluded, that even a volume of 8748 μm^3 is hardly representative for an anode substrate, but naturally the RVE size depends on the parameter under investigation. Between the biggest volumes, the parameters barely change, indicating that the largest volumes are representative volume elements for all parameters analysed here. However, in a further study volumes of similar size of the same substrate have to be analysed and the calculated parameters compared. The resulting standard deviation will give evidence if these volume size is sufficient to be a RVE. Analysing significantly larger volumes would also be of interest, but this will be very challenging as these volume sizes already constitute very large volume for FIB tomography.

The RVE size of the AFL is determined by using anode type D, as this anode has a much thicker AFL with 16 µm compared to 7 µm for anode type C. Thus the maximum cubic volume which could theoretically be analysed for type C is $7 \times 7 \times 7\ \mu m^3$ (L_{VE} of about 20). But as the electrolyte/anode interface is not straight at the reconstructed location of anode type C, it is not possible to analyse a cube of $(7\ \mu m)^3$ which only contains the AFL. For type D, however, it was possible to analyse a cubic volume of the AFL of up to $14 \times 14 \times 14\ \mu m^3$ (L_{VE} of over 40), the calculated parameters are plotted in Fig. A.3.

The material fractions change fast until a volume of ~10^3 particles and slightly afterwards. At about $L_{VE} = 34$ (about 1700 μm^3) the material fractions converge.

The surface area density converges much earlier. After a volume of 10^3 particles, the surface area density of YSZ and pores are barely changing, while the surface area density of nickel increases slightly before $L_{VE} = 30$. For volumes larger than 30^3 particles (or 1260 μm^3), the surface area densities remains constant.

In contrast, the tortuosities of nickel and pores hardly converge. Similar to the results of the substrate, they seem to converge only for the largest volumes of above $L_{VE} = 37$ (2200 μm^3). For smaller volumes, only the tortuosity of YSZ barely changes with the analysed volume size, showing thereby the lowest value. Similar to the pore tortuosity in the substrate, this can be explained with the high YSZ content of over 40 % in this layer. In contrast, the contents of pores and especially nickel are near the percolation threshold, which is the area where bottlenecks of transport paths occur and even small changes in the phase content can lead to significant changes in the tortuosity value.

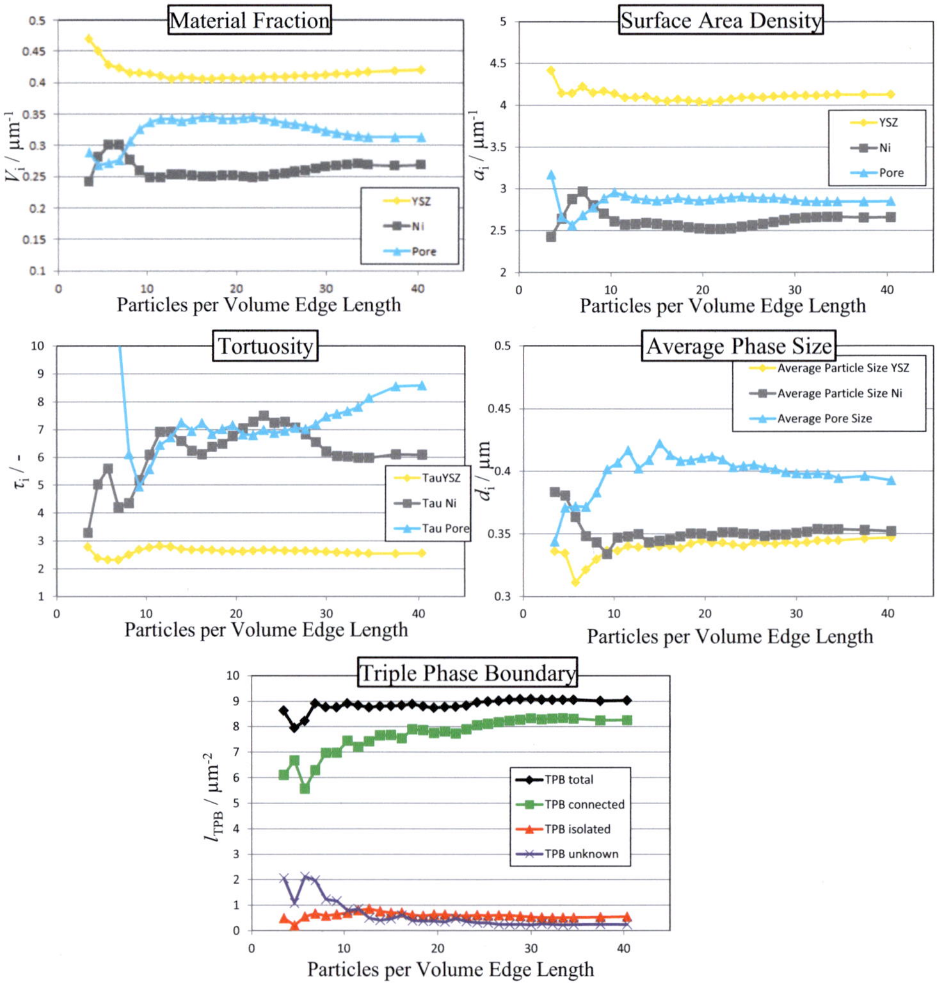

Figure A.3.: Calculated microstructural parameters for an increasing volume: material fractions, surface area densities, tortuosities, average phase sizes and triple-phase boundaries of the AFL of type D anode.

The courses of the average phase sizes are almost the same as for the substrate. Before $L_{VE} = 10$, the average phase size of all three phases are very sensitive to the analysed volume size. Afterwards, the average particle sizes of the two solid phases are relatively insensitive to the analysed volume size. In contrast, the average pore size decreases slightly with increasing volume and barely converges.

Looking at the triple-phase boundary density, especially the total amount of TPBs is unsensitive to the volume size, as the TPB density barely changes above $L_{VE} = 7$. However, the shares of percolating, unknown and isolated TPBs are slightly changing until L_{VE} of about 28. This means that adequate values for the TPB density can be obtained from volumes containing a minimum of 28^3 particles (1000 μm^3).

It can be concluded that (similar to the findings for the anode substrate) the RVE size for the AFL depends strongly on the parameter under investigation. Starting from volumes containing more than 10^3 particles (over 50 μm^3), most of the parameters are at least close to the correct values. However, only the largest volumes analysed (containing over 40^3 particles) can be seen as representative with respect to all parameters analysed here. As in the end all parameters are barely changing with increasing volume, it can be concluded that the values from the largest volumes are meaningful and representative, which holds for the anode substrate as well as the anode functional layer. Nevertheless, reconstructing a volume large enough to act as a RVE remains challenging for AFLs with only 7 µm thickness.

A.4. Results of Anode Quantification

This section provides the results from the microstructural characterisation of the anodes, whereby all four types are compared with each other. The functional layer and the substrate were analysed separately. The differences between the various electrodes will be discussed and evaluated with respect to their functionality. Please note that an exact comparison of the microstructures of the different types is not possible, as the ASCs have a different prehistory (see Table A.1): the anode microstructure can change during operation, depending on experimental conditions [131]. However, for the operation condition applied for the investigated cells of this work, only small changes are expected. Thus characteristic differences between the cells can still be detected, which enables the identification of the different optimization strategies of the manufacturers.

Table A.3.: Reconstructed volume sizes of all four anodes, as well as the analysed volume sizes for the AFL and anode substrate.

	Type A	Type B	Type C	Type D
Total volume / μm^3	1765	887	14094	7056
Analysed AFL volume / μm^3	972	328	1458	3906
Analysed substrate volume / μm^3	792	559	8748	2722

The reconstructed volumes of the double-layer anodes are listed in Table A.3 and shown in Fig. A.4. From the total volumes, sections only containing the AFL, and sections only containing the substrate are analysed. The volume sizes analysed for both layers are also listed in Table A.3. It can be seen that the analysed volumes of type B are small, while the volumes of type C and D are very large. As already discussed, different volume sizes can lead to different uncertainties for the calculated parameters. Especially for AFLs with a thickness of only 6 to 7 µm it is challenging to reconstruct large volumes. Nevertheless, for volumes containing more than 10^3 particles (which is the case for all analysed volumes), the parameters show at least an approximation of the right value, as shown in Section A.3.

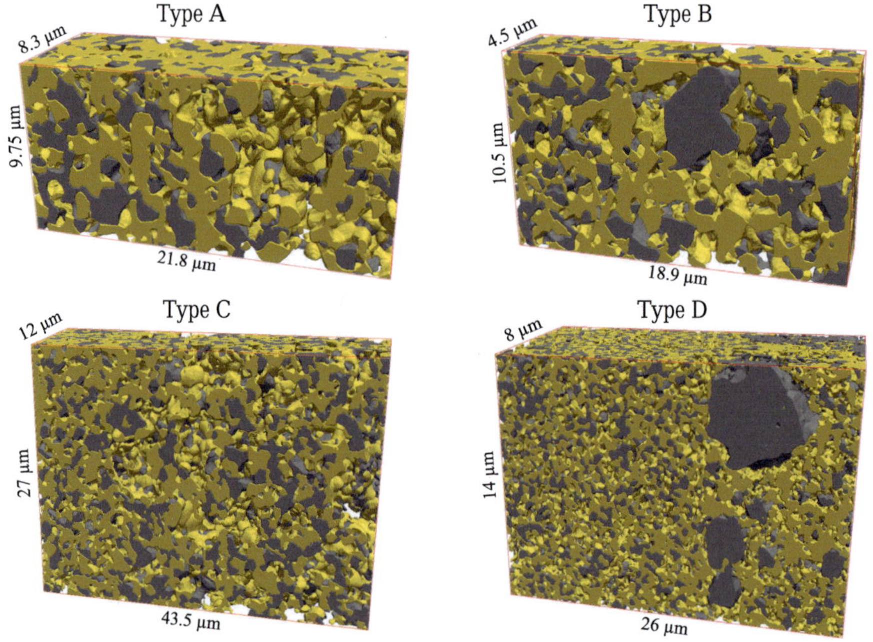

Figure A.4.: Total reconstructed volumes of the four types of double-layer anodes (not to scale). Ni grains appearing as grey, YSZ grains as yellow, and pores transparent.

Volume Fractions

The results of the volume fraction X_i of each individual phase i is given in Table A.4. The shares of the three phases are different for the anode substrate and the AFL. All four types have a larger amount of YSZ within the AFL compared to the substrate. On the other side, the porosity fraction inside the substrate is larger than inside the AFL, while the Ni fraction is only slightly lower for the substrate than for the AFL (except of type D, where the Ni content in the AFL is slightly higher). Looking at the main functionality of the two layers, it becomes obvious that the ionic conductivity inside the AFL is more important than inside the substrate. Inside the substrate the electronic conductivity and especially good transport paths for the gases are more important.

Within the AFL, only small differences between types A, B and C are observed. Only type D shows a much lager porosity with 31.1 % and a lower YSZ fraction with 41.7 %. Thus, the shares of all three phases are much closer to each other for type D compared to the other three types. However, besides a desired high ionic conductivity, a high YSZ fraction inside the AFL leads also to a good match of the thermal expansion coefficients between the electrolyte (consisting of YSZ) and the adjacent AFL (Ni/YSZ).

Table A.4.: Volume fractions X_i of the individual phases calculated separately for the AFL and the anode substrate, respectively.

Layer and Type	YSZ Fraction / %	Ni Fraction / %	Pore Fraction / %
AFL, Type A	48.4	29.4	22.3
AFL, Type B	49.3	28.5	22.2
AFL, Type C	49.3	30.2	20.5
AFL, Type D	41.7	27.2	31.1
Substrate, Type A	38.2	19.5	42.3
Substrate, Type B	41.4	24.6	34.0
Substrate, Type C	34.7	23.8	41.5
Substrate, Type D	26.6	34.3	39.2

As expected, the porosity inside the substrate is much higher than inside the AFL, but differs significantly among the four substrates. Types A and C show a high porosity with 42.3 % and 41.5 %, while especially type B has a much lower porosity fraction of only 34.0 %. Most probably, the high porosity should effectively compensate for a higher gas diffusion polarization in the 1500 μm thick substrates of types A and C, compared to the only 500 μm thick substrates of types B and D. For a structurally and chemically similar ASC as type A, an area specific resistance for the gas diffusion polarization of $\leq 20\,\mathrm{m\Omega cm^2}$ at a simulated fuel utilization of 65 % and for temperatures between 550 °C and 850 °C was reported in Ref. [115].

Phase Connectivity

As mentioned in Section 4.7.1, the connectivity of Ni, YSZ and pores is calculated using MATLAB. All clusters of voxels are labelled as *connected, unknown* or *isolated*. The shares of *connected* and *isolated* parts for the individual phases are listed in Table A.5, whereas the amount missing to 100 % are voxels labelled as *unknown*.

It can be seen that in general the connectivity is very high. The percentage of isolated pores within all substrates is very low with values between 0.10 % (type D) and 1.13 % (type C), but increases inside the AFL to 1.28 % for type D and 10.73 % for type C. This relatively high values of isolated pores inside the AFL (especially of type C) can be explained with

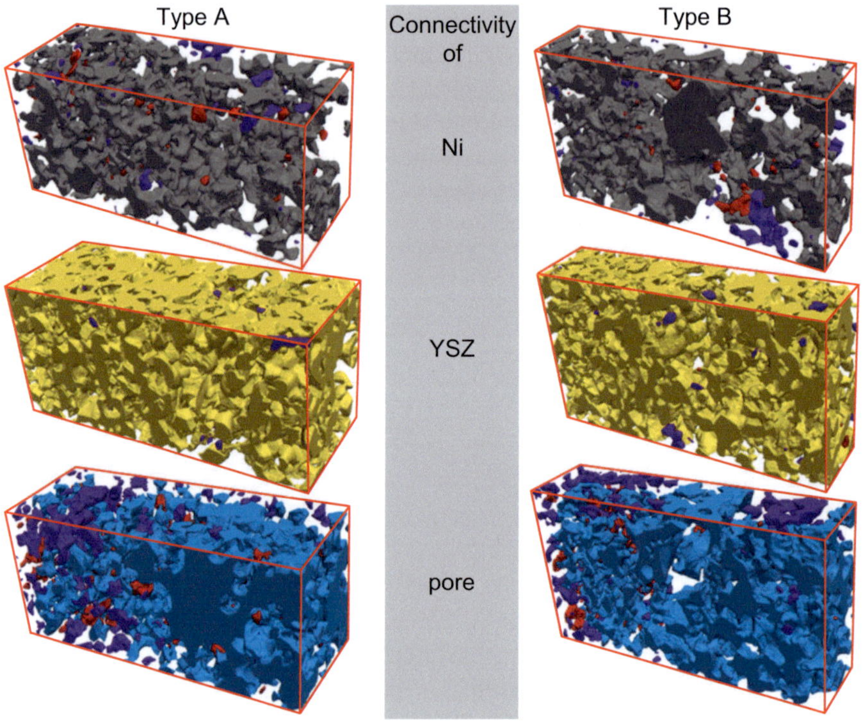

Figure A.5.: Results of the connectivity analysis for type A (left) and type B (right): red represents isolated parts and violet parts of unknown status, whereas the grey, yellow and blue (for Ni, YSZ and pores) represents parts that are connected to the electrolyte for the YSZ or to the current collector and fuel supply, respectively, for pores and nickel.

the low porosity fractions there (cf. Table A.4), because the probability that all pores are connected decreases. Only type D with its high porosity fraction shows an almost perfect connectivity of the pores inside the AFL. The high numbers of voxels with an *unknown* connectivity status of type A and B are mainly related to sample geometries, as in the image slicing direction the length is only 8.3 μm for type A and 4.5 μm for type B. The results of the connectivity analysis for type A and B are visualized in Fig. A.5.

The connectivity of YSZ is almost perfect with $\geq 99.16\,\%$ in both the AFL and the substrate. The connectivity of the nickel is slightly lower with $\geq 98.15\,\%$ inside the AFL and $\geq 93.27\,\%$ inside the substrate. Interestingly, the substrate with the highest amount of nickel has also the highest amount of isolated nickel (anode type D). However, if looking closely to the connectivity analysis of type D, it was observed that almost all *isolated* nickel particles are very small particles. It might be possible that the anode substrate is not well fabricated or that small nickel particles got isolated during operation. However, this is not very probable

Table A.5.: Percentage of *connected* (c) and *isolated* (i) voxels for the individual phases, calculated separately for the AFL and the anode substrate, respectively. The amount missing to 100 % are of *unknown* status.

Layer and Type	YSZ / %		Nickel / %		Porosity / %	
	c	i	c	i	c	i
AFL, Type A	99.88	0.07	98.15	1.13	76.79	5.95
AFL, Type B	99.94	0.04	98.50	0.40	77.61	8.37
AFL, Type C	99.95	0.02	99.48	0.28	83.88	10.73
AFL, Type D	99.97	0.01	98.94	0.85	97.63	1.28
Substrate, Type A	99.70	0.08	97.15	1.03	98.60	0.84
Substrate, Type B	99.30	0.16	94.78	1.06	96.92	0.39
Substrate, Type C	99.69	0.12	98.96	0.34	97.46	1.13
Substrate, Type D	99.16	0.40	93.27	5.30	99.82	0.10

as such phenomena (if existing) should also take place in the AFL. A more likely explanation is that this small particles located on or inside the YSZ matrix are only erroneously detected as nickel, but in fact they are YSZ. As already mentioned in Section A.1, the substrate of type D contains 3YSZ as the ionic conducting phase. In the consecutive SEM images, this 3YSZ is not of uniform color, but contains some slightly darker spots (cf. Fig. A.6). For some of this darker spots, it is hard to decide if they are 3YSZ, or if they are nickel. Hence it is possible that a non-negligible amount of YSZ was erroneously detected as nickel inside the substrate. In contrast, inside the AFL the YSZ is much more uniformly colored, as can be seen in Fig. A.6, which indicates that different compositions of YSZ are used for the AFL and the substrate.

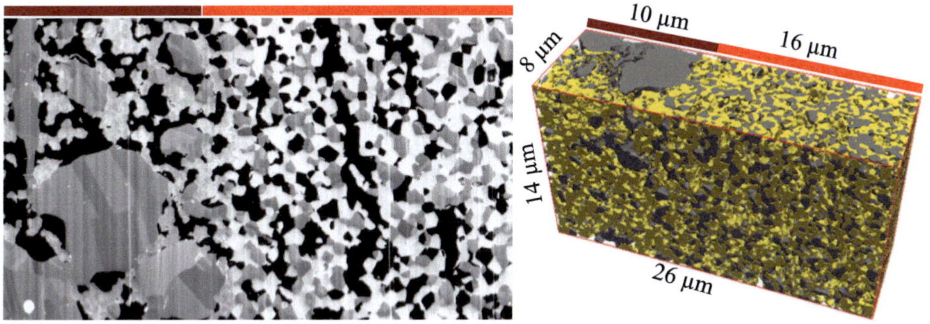

Figure A.6.: Section of one of the consecutive images of type D anode before image processing (left) and the 3D volume of the whole data set of type D after reconstruction (right).

Overall it can be concluded that the connectivity is very high, except of the porosity inside the AFL for the anodes type A, B and C.

Tortuosity

The tortuosity values of Table A.6 show that the tortuosity of nickel and YSZ is smaller inside the AFL than inside the substrate. The larger shares of nickel and YSZ inside the AFL and the smaller thickness of the layer are the main reasons for this. Especially the tortuosity of YSZ inside the AFL is very low, which is not surprising if looking at the large amounts of YSZ there. In contrast, the tortuosity of the pores inside the AFL show extremely high values, especially for type A and B. However, this values are not particularly meaningful, due to different reasons: inside the AFL the amount of pores is very low, especially on or near the interface to the electrolyte. Thus the calculated values are strongly affected by the analysed AFL-volume near the electrolyte. Calculating the tortuosity requires continuous transport paths. Due to the low amount of pores on or near the electrolyte, paths of the pores are ending or at least get constricted in the vicinity of the electrolyte. Thus, the tortuosity can change much depending on the location which is analysed of the AFL. A fair comparison of different AFLs is almost not possible, as it is hard to consider comparable volumes for several anodes.

Table A.6.: Tortuosity τ_i of the individual phases calculated separately for the AFL and the anode substrate, respectively.

Layer and Type	Tortuosity of YSZ	Tortuosity of Ni	Tortuosity of Pores
AFL, Type A	2.18	4.64	33.28
AFL, Type B	2.13	5.34	46.59
AFL, Type C	1.92	3.59	19.24
AFL, Type D	2.57	5.92	8.66
Substrate, Type A	3.19	9.25	4.53
Substrate, Type B	3.65	7.45	4.41
Substrate, Type C	3.49	5.18	3.03
Substrate, Type D	3.37	7.55	2.95

In contrast, comparing the tortuosity values of different substrates is much easier, due to its larger thickness and as the substrate is not adjacent to the dense electrolyte. The nickel phase has the largest tortuosity values for all four types of anodes (between 5.18 for type C and 9.25 for type A). This is not surprising for type A, B and C, as for these substrates the share of nickel is the lowest of all three phases. But also type D with its relatively high nickel fraction show a large tortuosity for the nickel with 7.55. However, due to the high nickel conductivity, these large tortuosity values are not a major problem.

The YSZ tortuosity values inside the substrate are very close to each other for all four types, ranging from 3.19 to 3.65. This is especially surprising for type D with a relative small amount of YSZ of only 26.6 % (34.7 to 41.4 % for type A, B and C). However, it shows the good connectivity of the YSZ matrix of the substrate for all four types of anodes.

The tortuosity values calculated for the pores differ slightly more than for the YSZ. But with values between 2.95 for type D and 4.53 for type A, they are still relatively small. This means that the transport of the gases inside the anode substrates are not hampered much by the structure.

Volume-Specific Surface Areas and Interfacial Areas

The results of the volume-specific surface areas and the interfacial surface areas calculated with the marching cubes algorithm are given in Table A.7. For the interfacial surface areas between pore and nickel ($A_{\mathrm{Ni/Pore}}$) and between pore and YSZ ($A_{\mathrm{YSZ/Pore}}$), the shares of *connected* (c), *unknown* (u) and *isolated* (i) surface areas are listed in Table A.8.

Table A.7.: Volume-specific surface areas of all phases for the AFL and the anode substrate of the four anode types, separately analysed for AFL and substrate.

Layer and Type	A_{YSZ} / μm^{-1}	A_{Ni} / μm^{-1}	A_{Pore} / μm^{-1}	$A_{\mathrm{Ni/Pore}}$ / μm^{-1}	$A_{\mathrm{Ni/YSZ}}$ / μm^{-1}	$A_{\mathrm{YSZ/Pore}}$ / μm^{-1}
AFL, Type A	2.33	1.62	1.68	0.49	1.14	1.20
AFL, Type B	3.11	2.17	2.28	0.67	1.50	1.61
AFL, Type C	2.43	1.64	1.67	0.44	1.20	1.23
AFL, Type D	4.12	2.68	2.84	0.70	1.99	2.14
Substrate, Type A	1.93	1.14	1.54	0.37	0.76	1.16
Substrate, Type B	2.17	1.34	1.82	0.49	0.84	1.33
Substrate, Type C	1.85	1.24	1.44	0.42	0.82	1.03
Substrate, Type D	2.46	1.71	2.86	1.06	0.65	1.80

The volume-specific surface area A_{Ni} and the interfacial surface area $A_{\mathrm{Ni/Pore}}$ are a measure for the accessible catalyst surface for oxidation of hydrogen and internal reforming reactions. It becomes evident, that type B with values of 2.17 μm^{-1} and 0.67 μm^{-1} and type D with values of 2.68 μm^{-1} and 0.70 μm^{-1} in the AFL are superior to type A and C, which shows 30 to 40 % lower values thereof. Type D substrate shows the highest values, while the difference among the other three anode substrates is less pronounced.

The values of the volume-specific surface areas A_i and the interfacial surface areas are higher for type B and D compared to type A and C. This suggests that the individual particles are

Table A.8.: Volume-specific interfacial surface areas of all phases for the AFL and the anode substrate of the four anode types, separately analysed for AFL and substrate.

Layer and Type	$A_{\mathrm{Ni/Pore}}$ / µm^{-1}			$A_{\mathrm{YSZ/Pore}}$ / µm^{-1}		
	c	u	i	c	u	i
AFL, Type A	0.35	0.08	0.06	0.86	0.22	0.11
AFL, Type B	0.50	0.08	0.09	1.26	0.17	0.18
AFL, Type C	0.38	0.01	0.04	0.97	0.05	0.21
AFL, Type D	0.68	0.01	0.01	2.05	0.03	0.06
Substrate, Type A	0.35	0.01	0.02	1.11	0.02	0.04
Substrate, Type B	0.46	0.02	0.01	1.26	0.06	0.01
Substrate, Type C	0.40	0.01	0.01	0.94	0.03	0.05
Substrate, Type D	1.05	0.00	0.00	1.80	0.00	0.01

smaller in type B and D anodes, and that the particles inside the AFL are smaller when compared to the anode substrate. The particle size is analysed in the following subsection.

The connectivity of the interfacial surface areas is very high, especially inside the substrates. Inside the AFL, only type D shows very low shares of isolated interfacial surface areas of between 1.4 and 2.8 % of the total interfacial surface areas. For type A, B and C, the shares of isolated surface areas are between 9 and 17 %.

Particle and Pore Size Distribution

The average and median values of the phase distributions are listed in Table A.9. As already mentioned, if both values are close to each other, it is very likely that the distribution has a Gaussian-shape. In almost all cases the median and average value are very similar, except for the nickel particles of type D and the pores of type A and C in the substrate (cf. Table A.9). The phase size distributions in Figs. A.7 and A.8 show that all other distributions can adequately be approximated by a Gaussian function.

However, significant differences exist between the different types of anodes, layers and phases. For example, the median pore size d_{pore} and the pore size distributions of the different anodes show significant differences, both in the AFL and the substrate. Type A and C show similar values especially for the substrate, but also for the AFL, even if the median pore size is about 12 % smaller for type C there. In contrast, the median pore sizes of type D is about 12 to 34 % smaller inside both layers compared to type A and C. Moreover, the phase size distributions of type B and D are much more narrow. Of particular note is that the pore size distributions of type A and C for the substrate are not of Gaussian shape, as both substrates contain some larger pores. Even if the number of these large pores is not high, they naturally contribute significantly to the pore volume. The pore size distribution in Fig.

Table A.9.: Phase sizes d_p (particle- and pore diameter) of Ni, YSZ and pores within the AFL and the substrate for all four types of anodes. The listed values represent the median and mean values of the calculated phase size distributions.

Layer and Type	YSZ / nm		Nickel / nm		Pore / nm	
	average	median	average	median	average	median
AFL, Type A	615	608	544	512	480	461
AFL, Type B	481	465	421	403	379	373
AFL, Type C	591	579	546	519	432	407
AFL, Type D	340	339	344	336	379	365
Substrate, Type A	618	615	554	523	710	595
Substrate, Type B	587	569	508	465	563	534
Substrate, Type C	592	592	532	508	728	592
Substrate, Type D	329	310	403	255	410	392

A.8 shows that the pores becomes as large as 2.5 µm in the substrate of type A and C, but only as large as 1 to 1.5 µm for type D and B, respectively. In agreement with the inspection of SEM images (see e.g. Fig. 3.2), the pore sizes are sufficiently large for low-loss gas diffusion from the gas channel to the AFL.

The same trends can also be seen for the two solid phases. Nickel and YSZ particles of type A and C are similarly large. In contrast, the median particle sizes are slightly smaller (4 and 24 %) for type B and significantly smaller (35 and 52 %) for type D. Furthermore, for type A and C it is anticipated that the powder quality of Ni and YSZ is the same for AFL and anode substrate, as e.g. for type A the median particle size for YSZ is 608 nm and 615 nm, and for Ni 512 nm and 523 nm. For type B, Ni as well as YSZ particles inside the AFL are almost 20 % smaller compared to the substrate. For type D, especially the particle size distributions of nickel show big differences between the AFL and substrate. While in the AFL the distribution has a narrow Gaussian shape, the substrate contains some extremely large particles (up to 2.5 µm). But as much more smaller particles are detected, the median particle size is even smaller in the substrate, even if the average is larger, which clearly indicates the differces between the layers. However, the smaller average particle sizes of type B and D in the AFL resulted in higher surface areas compared to the values for type A and C, as discussed above. This will most probably also hold for the TPB densities, which will be investigated in the following subsection.

Triple-Phase Boundary (TPB)

The triple-phase boundary is of particular importance for the electrochemical performance of the anode, as the oxidation reaction occurs at the TPB. As desired, the $l_{TPB,total}$ is highest

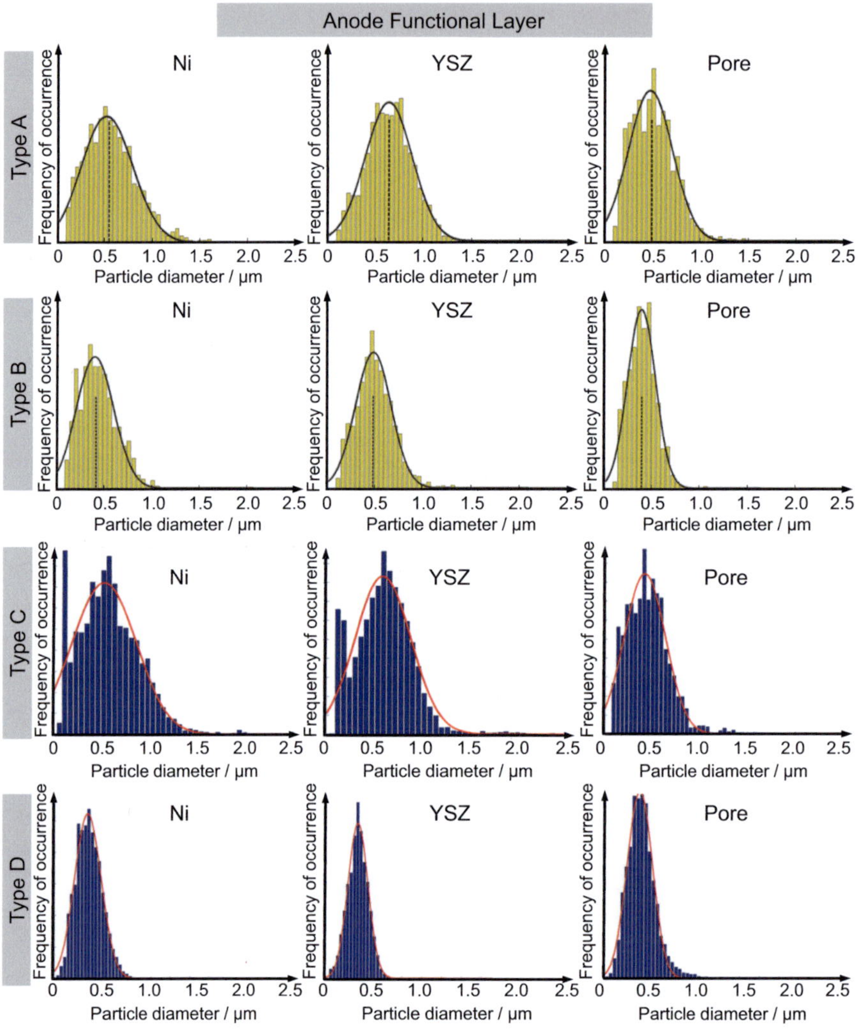

Figure A.7.: Phase size distributions of the AFLs calculated for all four types of anodes. The dotted lines indicate the median values of the given distributions (see Table A.9), while the curved lines approximate the distribution by Gaussian functions.

inside the AFL for all four ASCs (cf. Table A.10). Inside the substrate, the values for $l_{\mathrm{TPB,total}}$ are about 25 to 50 % lower. However, the specific values for the different anode types are differing very much. The total TPB density of type A and C are very similar with 2.56 µm^{-2} and 2.75 µm^{-2}, while $l_{\mathrm{TPB,total}}$ of type B is almost twice as large with 4.36 µm^{-2}.

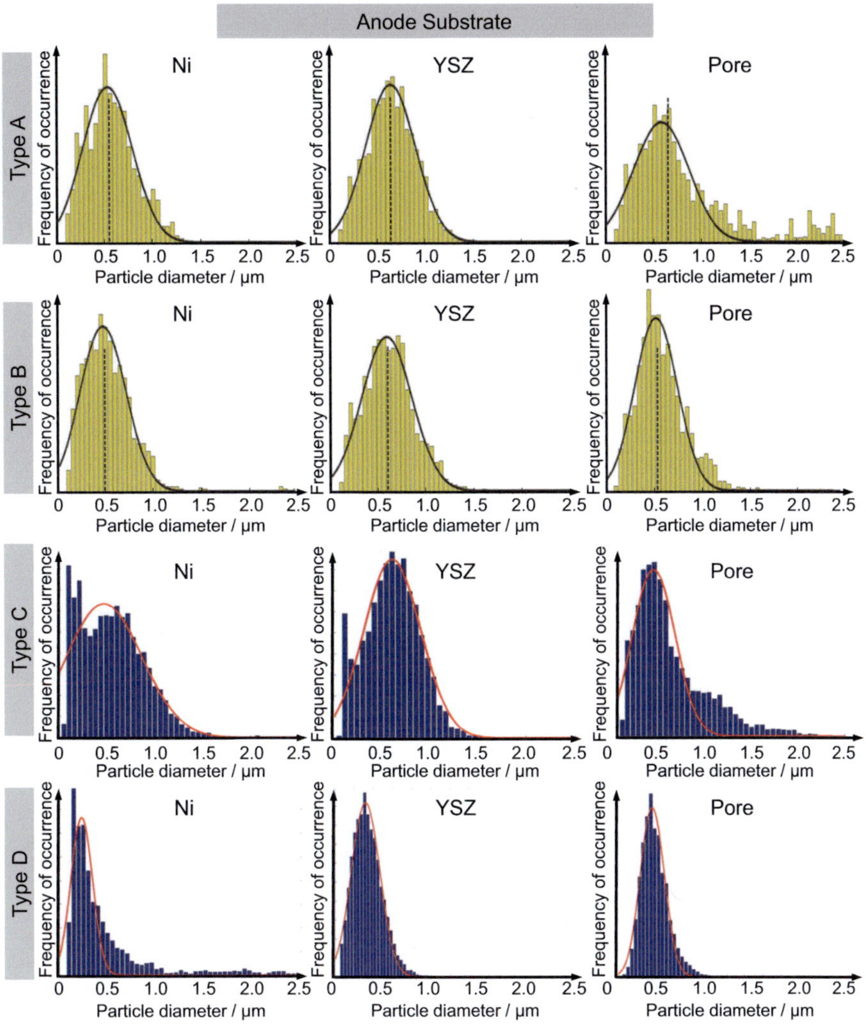

Figure A.8.: Phase size distributions of the substrates calculated for all four types of anodes. The dotted lines indicate the median values of the given distributions (see Table A.9), while the curved lines approximate the distributions by Gaussian functions.

Type D shows by far the largest triple phase boundary density with 6.22 μm^{-2} inside the AFL. One of the main reasons for the high values are the small particle and pore sizes of type D compared to the other three anode types (cf. Table A.9). Another reason is the volume fractions of the individual phases: while the YSZ fraction is more than double as large as the pore fraction for type A, B and C, the shares of all three phases are much more balanced for type D. In case that all three phases do have the same volume fraction, the TPB density would theoretically reach its maximum.

Table A.10.: Calculated triple-phase boundary density (l_{TPB}) for the AFL and the substrate. The densities were calculated for all TPBs ($l_{TPB,total}$), for TPBs where the connectivity-status of at least one phase is unknown ($l_{TPB,unknown}$) and for inactive TPBs, where at least one phase is isolated ($l_{TPB,isolated}$).

Layer and Type	$l_{TPB,total}$ / μm^{-2}	$l_{TPB,connected}$ / μm^{-2}	$l_{TPB,unknown}$ / μm^{-2}	$l_{TPB,isolated}$ / μm^{-2}
AFL, Type A	2.56	1.56	0.51	0.49
AFL, Type B	4.36	2.70	0.76	0.91
AFL, Type C	2.75	2.06	0.21	0.48
AFL, Type D	6.22	5.68	0.15	0.39
Substrate, Type A	1.76	1.43	0.12	0.22
Substrate, Type B	2.14	1.60	0.34	0.20
Substrate, Type C	1.91	1.61	0.10	0.19
Substrate, Type D	4.63	3.07	0.25	1.32

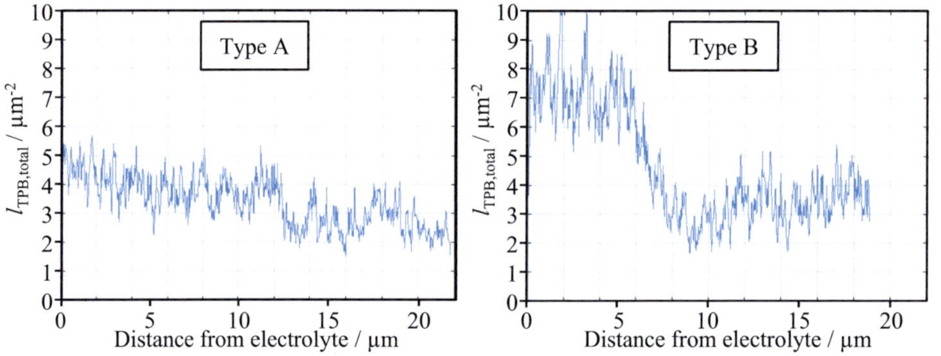

Figure A.9.: Distribution of total TPB density $l_{TPB,total}$ calculated from the AFL/electrolyte interface towards the current collector/gas channel, thus showing the decrease of $l_{TPB,total}$ from the AFL into the substrate. Note: $l_{TPB,total}$ is calculated directly from the voxel mesh here by a simple edge summation.

Figure A.9 compares $l_{TPB,total}$ of type A and type B double layer anodes as a function of the distance to the electrolyte. For this, the $l_{TPB,total}$ is calculated for each individual slice directly by edge summation (cf. Section A.2), starting from the AFL/electrolyte interface to the interface AFL/substrate and towards the direction of the current collector/gas channel. This analysis discloses a large difference in AFL extension between both types of anodes, as already indicated in the SEM images. Type B shows a clear drop of $l_{TPB,total}$ in a distance of 6 to 9 µm from the electrolyte, which is the interface AFL/substrate. In contrast, $l_{\mathrm{TPB,total}}$ gives only a slightly decreasing trend for type A. The interface AFL/substrate is set at a distance of 12 µm from the electrolyte, but could possibly extend more. Interestingly enough, the lateral extension of the electrochemical active volume was already determined in Ref. [213], to be increasing from 10 µm at 950 °C to 14 µm at 750 °C for a chemically similar, but structurally different type of anode. Hopefully, a critical examination of the relationship between performance and TPB length becomes tangible with the progress presented here.

B. Transformation of ALS Model

In the original ALS model [60], the transport of oxygen ions is described in terms of oxygen vacancies as the mobile species, using the vacancy diffusion coefficient D_v and the associated surface exchange coefficient k together with the thermodynamic enhancement factor γ_{v} for vacancies (denoted as A in Ref. [60] and defined as $\gamma_{\mathrm{v}} = -0.5 \cdot \partial \ln(pO_2) \setminus \partial \ln(c_{\mathrm{v}})$). However, to describe mixed conducting materials as cathodes in SOFCs, the chemical diffusion coefficient D^{δ} and chemical surface exchange coefficient k^{δ} are used, together with the thermodynamic factor (defined as $\gamma_{\mathrm{o}} = 0.5 \cdot \partial \ln(pO_2) \setminus \partial \ln(c_{\mathrm{o}})$). Therefore, a transformation is necessary, as described in the following.

Surface Exchange Coefficient and Relaxation Time

In Ref. [60], where the original ALS model is presented, Equation 16 describes:

$$t_{chem} = \frac{c_v(1-\varepsilon)}{Aar_0(\alpha_f+\alpha_b)}, \tag{6.3}$$

where c_v is the vacancy concentration, ε the porosity, a is the surface area density, A is the thermodynamic factor (denoted as γ_{v} within this thesis and in the following), r_0 is the exchange neutral flux density (in analogy to exchange current density) and α_{f} and α_{b} are constants of the order of unity that depend on the specific mechanism of the exchange reaction [60]. However, the last part can be described using the surface coefficient k and the concentration of oxygen lattice sites in the mixed conductor c_{mc}: $r_0(\alpha_{\mathrm{f}}+\alpha_{\mathrm{b}}) = kc_{\mathrm{mc}}$ (Eq. 23 in [60]). With this, Eq. (6.3) can be written as

$$t_{\mathrm{chem}} = \frac{c_{\mathrm{v}}(1-\varepsilon)}{\gamma_{\mathrm{v}} a c_{\mathrm{mc}} k}. \tag{6.4}$$

From [3], Eqs. 2-41 and 2-42, it can be followed that $c_v = \frac{\delta}{3-\delta} c_o$, where c_o is the concentration of oxygen ions. Hence,

$$t_{chem} = \frac{\frac{\delta}{3-\delta} c_o (1-\varepsilon)}{\gamma_v a c_{mc} k}. \tag{6.5}$$

Equation 3.37 from [26] can be used to transform the thermodynamic factor of vacancies (γ_v) to that of oxygen concentration (γ_o): $\gamma_v = \frac{c_{mc} - c_o}{c_o} \gamma_o$. It should be noted, that the γ_v – in contrast to γ_o – shows a strong dependency from the oxygen partial pressure pO_2, due to the dependency of c_o on pO_2. As $c_o = \frac{3-\delta}{3} c_{mc}$ (Eqs. 2.42 and 2.43 from [3]), it follows that $\gamma_v = \frac{c_{mc} - \frac{3-\delta}{3} c_{mc}}{\frac{3-\delta}{3} c_{mc}} \gamma_o = \frac{3-(3-\delta)}{3-\delta} \gamma_o = \frac{\delta}{3-\delta} \gamma_o$ and hence

$$t_{chem} = \frac{\frac{\delta}{3-\delta} c_o (1-\varepsilon)}{\frac{\delta}{3-\delta} \gamma_o a c_{mc} k}, \tag{6.6}$$

or equivalently after canceling

$$t_{chem} = \frac{c_o (1-\varepsilon)}{\gamma_o a c_{mc} k}. \tag{6.7}$$

As $k^\delta = \gamma_o k$ ([66], Eq. 7), this finally leads to

$$t_{chem} = \frac{c_o (1-\varepsilon)}{a c_{mc} k^\delta}. \tag{6.8}$$

Hence, the chemical surface exchange coefficient can be calculated with the following formula:

$$k^\delta = \frac{c_o (1-\varepsilon)}{a c_{mc} t_{chem}}. \tag{6.9}$$

Oxygen Ion Diffusion Coefficient and Characteristic Resistance

In Ref. [60] Equation 18 describes:

$$R_{chem} = \left(\frac{RT}{2F^2}\right) \sqrt{\frac{\tau}{(1-\varepsilon) c_v D_v a r_0 (\alpha_f + \alpha_b)}}. \tag{6.10}$$

With Eq. 23 from [60] ($r_0(\alpha_f + \alpha_b) = k c_{mc}$), this can be written as

$$R_{chem} = \left(\frac{RT}{2F^2}\right) \sqrt{\frac{\tau}{(1-\varepsilon) c_v D_v a c_{mc} k}}. \tag{6.11}$$

Using Eq. 24 from [60] ($D^* = fD_\mathrm{v}\frac{c_\mathrm{v}}{c_\mathrm{mc}}$) with the same assumption used there, that f is unity ($f = 1$), this leads to

$$R_\mathrm{chem} = \left(\frac{RT}{2F^2}\right)\sqrt{\frac{\tau}{(1-\varepsilon)c_\mathrm{v}\frac{c_\mathrm{mc}}{c_\mathrm{v}}D^* a c_\mathrm{mc} k}}, \tag{6.12}$$

or after reducing the equation

$$R_\mathrm{chem} = \left(\frac{RT}{2F^2}\right)\sqrt{\frac{\tau}{(1-\varepsilon)D^* a c_\mathrm{mc}^2 k}}. \tag{6.13}$$

Using Eq. 2.10 from [26] or Eq. 6 from [67], respectively ($D^\delta = \gamma_\mathrm{o} D^*$ and $k^\delta = \gamma_\mathrm{o} k$), this leads to

$$R_\mathrm{chem} = \left(\frac{RT}{2F^2}\right)\sqrt{\frac{\tau}{(1-\varepsilon)\frac{D^\delta}{\gamma_\mathrm{o}} a (c_\mathrm{mc})^2 \frac{k^\delta}{\gamma_\mathrm{o}}}}, \tag{6.14}$$

which is equal to

$$R_\mathrm{chem} = \left(\frac{RT}{2F^2}\right)\sqrt{\frac{\tau\gamma_\mathrm{o}^2}{(1-\varepsilon)D^\delta a (c_\mathrm{mc})^2 k^\delta}}. \tag{6.15}$$

Hence, the chemical oxygen ion diffusion coefficient (in case of symmetrical cathode) can be calculated with the following equation:

$$D^\delta = \left(\frac{RT}{2F^2}\right)^2 \frac{\tau\gamma_\mathrm{o}^2}{(1-\varepsilon)a(c_\mathrm{mc})^2 k^\delta R_\mathrm{chem}^2}. \tag{6.16}$$

Note, that Eq. (6.15) expresses the characteristic resistance in case of a symmetrical cathode. In case of only one cathode (as it is the case for our anode-supported cells, where the cathode resistance is extracted from the full cell resistance), the characteristic resistance in the formula halves (e.g. this was mistakenly not considered in Ref. [2] while calculating D^δ, wherefore the values published there have to be divided by a factor of four in order to correct them). However, in case of only one cathode the diffusion coefficient must be calculated with the following equation:

$$D^\delta = \left(\frac{RT}{4F^2}\right)^2 \frac{\tau\gamma_\mathrm{o}^2}{(1-\varepsilon)a(c_\mathrm{mc})^2 k^\delta R_\mathrm{chem}^2}, \tag{6.17}$$

or by applying Eq. (6.8) by

$$D^\delta = \left(\frac{RT}{4F^2}\right)^2 \frac{\tau\gamma_\mathrm{o}^2 t_{chem}}{(1-\varepsilon)^2 c_\mathrm{mc} c_\mathrm{o} R_\mathrm{chem}^2}. \tag{6.18}$$

Alternative Translation for Oxygen Ion Diffusion Coefficient

There is an alternative formulation to transform D_v to D^δ, which for the sake of completeness will be shown in the following. Note, that the result will be slightly different, also this has almost no effect for the investigated materials LSCF and LSC. Using Eqs. 2-41 and 2-42 from [3], it can be conclude that $c_v = \frac{\delta}{3-\delta} c_o$. Analog to the discussion before Eq. (6.6) (using $c_o = \frac{3-\delta}{3} c_{mc}$ (Eqs. 2.42 and 2.43 from [3]) with Eq. 3.37 from [26]), it follows that $\gamma_v = \frac{\delta}{3-\delta}\gamma_o$. This can be inserted into the Equation $D_v = \frac{\tilde{D}}{A}$ from [230] (note, that here $D^\delta = \tilde{D}$ and $\gamma_v = A$) and hence $D_v = \frac{D^\delta}{\gamma_v} = \frac{3-\delta}{\delta}\frac{1}{\gamma_o} D^\delta$. Putting this information into Eq. (6.11) leads to

$$R_{chem} = \left(\frac{RT}{2F^2}\right)\sqrt{\frac{\tau}{(1-\varepsilon)\frac{\delta}{3-\delta}c_o \frac{3-\delta}{\delta}\frac{D^\delta}{\gamma_o} a c_{mc} k}}. \tag{6.19}$$

As $k^\delta = \gamma_o k$, this results in

$$R_{chem} = \left(\frac{RT}{2F^2}\right)\sqrt{\frac{\tau}{(1-\varepsilon)c_o \frac{D^\delta}{\gamma_o} a c_{mc} \frac{k^\delta}{\gamma_o}}} \tag{6.20}$$

and hence to

$$R_{chem} = \left(\frac{RT}{2F^2}\right)\sqrt{\frac{\tau\gamma_o^2}{(1-\varepsilon)c_o D^\delta a c_{mc} k^\delta}}. \tag{6.21}$$

And finally by applying Eq. (6.8) and taking into account that we have only one cathode:

$$D^\delta = \left(\frac{RT}{4F^2}\right)^2 \frac{\tau\gamma_o^2 t_{chem}}{(1-\varepsilon)^2 c_o^2 R_{chem}^2}. \tag{6.22}$$

The difference between Eq. (6.22) and (6.17) (c_o^2 instead of $c_{mc}c_o$) is due to the fact, that in Eq. 24 from [60], c_{mc} (most probably for reasons of simplification) is used instead of c_o. However, the difference is very small for LSC and LSCF, as $c_o = \frac{3-\delta}{3} c_{mc}$ with $3-\delta \approx 2.9$ for these materials.

Characteristic Length or Size of Active Region

In Ref. [60] Equation 15, the characteristic distance (denoted as δ there, but in order to avoid confusion with the nonstoichiometry, the characteristic distance will be denoted as l_δ in accordance with later publications [52] here), which describes the size of the active region [230], is defined by

$$l_\delta = \sqrt{\frac{c_v D_v (1-\varepsilon)}{a r_0 (\alpha_f + \alpha_b)\tau}}. \tag{6.23}$$

Analogous to above, this leads to

$$l_\delta = \sqrt{\frac{\frac{\delta}{3-\delta} c_\mathrm{O} \frac{3-\delta}{\delta} \frac{D^\delta}{\gamma_\mathrm{o}} (1-\varepsilon)}{akc_\mathrm{mc}\tau}} \tag{6.24}$$

and with $k^\delta = k\gamma_\mathrm{o}$

$$l_\delta = \sqrt{\frac{c_\mathrm{O} D^\delta (1-\varepsilon)}{ak^\delta c_\mathrm{mc}\tau}}. \tag{6.25}$$

C. Sintering

The following part is mainly based on the Refs. [225] and [231], where especially in [225] more details about sintering can be found. Sintering is a process for manufacturing ceramics (and metals), which is typically used for manufacturing the electrodes of SOFCs. In this process, fine grained ceramic (and/or metal) particles which are close together (e.g. as resin coated powders) are heated. Thereby the heating temperature is typically below the melting point temperature, but above about $2/3$ of the melting point temperature of the involved materials. Influenced by the heat, the grains start to grow and fusing together. Thus, sintering leads to a significant increase in the density (hence a decrease of porosity) and strength. The growth takes place at the atomic level by atoms diffusing across the boundaries of the particles, fusing the particles together and creating bigger particles. Driving force of this process is the minimization of the free surface energy. As the free surface energy is decreasing with increasing particle-diameter, small particles will fuse to bigger ones. Factors influencing the sintering behavior are sintering time and temperature, surrounding atmosphere, particle sizes and the sintering tendency of the material. During sintering of different materials in two successive layers, inter-diffusion is almost unavoidable, which can result in the formation of (unwanted) secondary phases.

D. Supervised Diploma Theses and Study Projects

- Andreas Häffelin: *Entwicklung eines 3D FEM Modells zur Untersuchung der Verlustmechanismen in mischleitenden Elektroden* (Supervised together with Moses Ender). Diploma thesis, 02.01.2011 - 24.01.2012.
- Gina Kristin Steinke: *Elektrochemische Charakterisierung mikrostrukturell modifizierter Kathoden für die SOFC* (Supervised together with André Leonide). Study project, 01.02.2011 - 30.11.2011.

- Ingo Rotscholl: *Rekonstruktion einer zweiphasigen Elektrode mittels FIB/REM* (Supervised together with Moses Ender). Bachelor thesis, 20.04.2011 - 30.10.2011.
- Andreas Messner: *Investigation of Novel Catalytic Active SOFC Anodes* (external; Northwestern University, USA). Bachelor thesis, 15.06.2012 - 30.11.2012.
- Florian Wankmüller: *Einfluss von Fremdphasen an der Kathoden/Elektrolyt Grenzfläche – Untersuchung und Modellentwicklung mittels FIB-Tomographie.* Master thesis (Supervised together with Julian Szász), 01.06.2014-21.05.2015.

E. Publications

- C. Endler-Schuck, J. Joos, C. Niedrig, A. Weber and E. Ivers-Tiffée, "The chemical oxygen surface exchange and bulk diffusion coefficient determined by impedance spectroscopy of porous $La_{0.58}Sr_{0.4}Co_{0.2}Fe_{0.8}O_{3-d}$ (LSCF) cathodes", Solid State Ionics 269, pp. 67-79 (2015).
- H. Geisler, A. Kromp, J. Joos, A. Weber and E. Ivers-Tiffée, "Stationary 2D FEM Model Framework for SOFC Stack Performance Prediction", ECS Trans. 68, pp. 3043-3050 (2015).
- C. Endler-Schuck, J. Joos, A. Weber and E. Ivers-Tiffée, "Durability Aspects of MIEC Cathodes", Proceedings of the 11th European Solid Oxide Fuel Cell Forum, Chapter 12, pp. 105-112 (2014).
- A. Messner, J. Joos, A. Weber and E. Ivers-Tiffée, "Custom Tailoring High-Performance MIEC Cathodes", Proceedings of the 11th European Solid Oxide Fuel Cell Forum, Chapter 10, pp. 157-163 (2014).
- M. Ender, J. Joos, A. Weber and E. Ivers-Tiffée, "Anode microstructures from high-energy and high-power lithium-ion cylindrical cells obtained by X-ray nano-tomography", J. of Power Sources 269, pp. 912-919 (2014).
- J. Joos, M. Ender, I. Rotscholl, N. Menzler, and E. Ivers-Tiffée, "Quantification of Double-Layer Ni/YSZ Fuel Cell Anodes from Focused Ion Beam Tomography Data", J. of Power Sources 246, pp. 819-830 (2014).
- A. Häffelin, J. Joos, A. Weber and E. Ivers-Tiffée, "Three-Dimensional Performance Simulation of SOFC Anodes Using FIB-Tomography Reconstructions", ECS Trans. 57 (1), pp. 2563-2572 (2013).
- A. Häffelin, J. Joos, M. Ender, A. Weber and E. Ivers-Tiffée, "Time-Dependent 3D Impedance Model of Mixed-Conducting Solid Oxide Fuel Cell Cathodes", J. Electrochem. Soc. 160 (8), pp. F867-F876 (2013).

- J. Joos, M. Ender, T. Carraro, A. Weber and E. Ivers-Tiffée, “Representative Volume Element Size for Accurate Solid Oxide Fuel Cell Cathode Reconstructions from Focused Ion Beam Tomography Data“, Electrochimica Acta 82, pp. 268-276 (2012).
- J. Joos, M. Ender, I. Rotscholl, N. Menzler, A. Weber and E. Ivers-Tiffée, “Quantification of Ni/YSZ-Anode Microstructure Parameters derived from FIB-tomography“, in F. Lefebvre-Joud (Ed.), Proceedings of 10th European SOFC Forum, Chapter 14, pp. 78-89 (2012).
- A. Häffelin, J. Joos, J. Hayd, A. Weber and E. Ivers-Tiffée, “3D FEM Impedance Model for LSCF-Cathodes“, in F. Lefebvre-Joud (Ed.), Proceedings of 10th European SOFC Forum, Chapter 17, pp. 45-55 (2012).
- T. Carraro, J. Joos, B. Rüger, A. Weber and E. Ivers-Tiffée, “3D Finite Element Model for Reconstructed Mixed-Conducting Cathodes: I. Performance quantification“, Electrochimica Acta 77, pp. 315-323 (2012).
- T. Carraro, J. Joos, B. Rüger, A. Weber and E. Ivers-Tiffée, “3D Finite Element Model for Reconstructed Mixed-Conducting Cathodes: II. Parameter sensitivity analysis“, Electrochimica Acta 77, pp. 309-314 (2012).
- T. Carraro and J. Joos, “Parameter Estimation for a Reconstructed SOFC Mixed-Conducting LSCF-Cathode“, in H. G. Bock, T. Carraro, W. Jäger, S. Körkel, R. Rannacher, and J. P. Schlöder (Eds.), Model Based Parameter Estimation. Theory and applications, Heidelberg: Springer, pp. 267-285 (2013).
- M. Ender, J. Joos, T. Carraro and E. Ivers-Tiffée, “Quantitative Characterization of LiFePO4 Cathodes Reconstructed by FIB/SEM Tomography“, J. Electrochem. Soc. 159, pp. A972-A980 (2012).
- A. Häffelin, J. Joos, M. Ender, A. Weber and E. Ivers-Tiffée, “Transient 3D Impedance Model for mixed-conducting SOFC Cathodes“, ECS Trans. 45 (1), pp. 313-325 (2012).
- J. Joos, T. Carraro, M. Ender, B. Rüger, A. Weber and E. Ivers-Tiffée, “Detailed Microstructure Analysis and 3D Simulations of Porous Electrodes“, ECS Trans. 35, pp. 2357-2368 (2011).
- A. Weber, A. Utz, J. Joos, E. Ivers-Tiffée, H. Störmer, D. Gerthsen, V. Yurkiv, H.-R. Volpp and W. G. Bessler, “Electrooxidation of Reformate Gases at Model Anodes“, ECS Trans. 35, pp. 1513-1528 (2011).
- A. Utz, J. Joos, A. Weber and E. Ivers-Tiffée, “Electrochemical Oxidation at SOFC Anodes: Comparison of Patterned Ni Anodes and Technical Ni/8YSZ Cermet Anodes“, ECS Trans. 35, pp. 1669-1682 (2011).

- M. Ender, J. Joos, T. Carraro and E. Ivers-Tiffée, “Three-Dimensional Reconstruction of a Composite Cathode for Lithium-Ion Cells“, Electrochemistry Communications 13, pp. 166-168 (2011).
- J. Joos, T. Carraro, A. Weber and E. Ivers-Tiffée, “Reconstruction of porous electrodes by FIB/SEM for detailed microstructure modeling“, J. Power Sources 196, pp. 7302-7307 (2011).
- J. Joos, B. Rüger, A. Weber, T. Carraro and E. Ivers-Tiffée, “3D Reconstruction of Porous Electrodes and Microstructure Modelling“, in DECHEMA (Ed.), First International Conference on Materials for Energy, Extended Abstracts - Book A, Karlsruhe, Germany, pp. 42-44 (2010).
- J. Joos, T. Carraro, A. Weber and E. Ivers-Tiffée, “Reconstruction of Porous Electrodes by FIB/SEM for Detailed Microstructure Modeling“, in J. T. S. Irvine and U. Bossel (Eds.), Proceedings of the 9th European Solid Oxide Fuel Cell Forum, pp. 6/75-6/81 (2010).
- J. Joos, B. Rüger, T. Carraro, A. Weber and E. Ivers-Tiffée, “Electrode Reconstruction by FIB/SEM and Microstructure Modeling“, ECS Trans. 28, pp. 81-91 (2010).
- B. Rüger, J. Joos, T. Carraro, A. Weber and E. Ivers-Tiffée, “3D Electrode Microstructure Reconstruction and Modelling“, ECS Trans. 25, pp. 1211-1220 (2009).
- B. Rüger, T. Carraro, J. Joos, V. Heuveline and E. Ivers-Tiffée, “A Numerical Method for the Optimization of MIEC Cathode Microstructures“, in R. Steinberger-Wilckens and U. Bossel (Eds.), Proceedings of the 8th European Solid Oxide Fuel Cell Forum, p. A0616 (2008).

F. Conference Contributions

- H. Geisler, A. Kromp, J. Joos, A. Weber and E. Ivers-Tiffée, Stationary 2D FEM Model Framework for SOFC Stack Performance Prediction, SOFC XIV (Glasgow , Scotland), 26.07. - 31.07.2015.
- C. Endler-Schuck, J. Joos, A. Weber and E. Ivers-Tiffée, Durability Aspects of MIEC Cathodes, 11th European SOFC Forum (Lucerne, Switzerland), 01.07. - 04.07.2014.
- A. Messner, J. Joos, A. Weber and E. Ivers-Tiffée, Custom Tailoring High-Performance MIEC Cathodes, 11th European SOFC Forum (Lucerne, Switzerland), 01.07. - 04.07.2014.
- J. Joos, C. Endler-Schuck, A. Weber, E. Ivers-Tiffée, Durability Analysis of LSCF Perovskites for Intermediate Temperature Solid Oxide Fuel Cells, 2014 MRS Spring Meeting & Exhibit (San Francisco, USA), 21.04. - 25.04.2014.

- J. Joos, A. Häffelin, M. Ender, A. Weber, E. Ivers-Tiffée, Performance Simulation of SOFC Anodes using a 3D FEM Model and Microstructure Data from FIB Tomography, 2014 MRS Spring Meeting & Exhibit (San Francisco, USA), 21.04. - 25.04.2014.
- A. Häffelin, J. Joos, A. Weber, E. Ivers-Tiffée, Three-Dimensional Performance Simulation of SOFC-Anodes Using FIB-Tomography Reconstructions, 13th International Symposium on Solid Oxide Fuel Cell (Okinawa, Japan), 06.10. - 11.10.2013.
- J. Joos, M. Ender, T. Carraro, A. Häffelin, A. Weber, E. Ivers-Tiffée, Modeling and Reconstruction of Mixed Conducting Cathodes for Solid Oxide Fuel Cells, Solid State Electrochemistry for Energy Storage and Conversion 2013 (Heidelberg, Germany), 22.07. - 24.07.2013.
- M. Ender, J. Joos, T. Carraro, A. Weber, E. Ivers-Tiffée, Microstructure modeling of porous electrodes for Lithium-ion batteries, Solid State Electrochemistry for Energy Storage and Conversion 2013 (Heidelberg, Germany), 22.07. - 24.07.2013.
- S. Wetterauer, T. Carraro, J. Joos, M. Ender, Extended Finite Elements for the Simulation of 3D Microstructures, Solid State Electrochemistry for Energy Storage and Conversion 2013 (Heidelberg, Germany), 22.07. - 24.07.2013.
- A. Häffelin, J. Joos, M. Ender, A. Weber, E. Ivers-Tiffée, Time-dependent 3D FEM Impedance Model of Mixed-Conducting Solid Oxide Fuel Cell Cathodes, Solid State Electrochemistry for Energy Storage and Conversion 2013 (Heidelberg, Germany), 22.07 - 24.07.2013.
- J. Joos, M. Ender, T. Carraro, A. Häffelin, E. Ivers-Tiffée, Quantification of SOFC Electrode Microstructure Characteristics, SSI-19 (Kyoto, Japan), 02.06 - 07.06.2013.
- M. Ender, J. Joos, A. Weber, E. Ivers-Tiffée, 3D Microstructure Analysis of Lithium-ion Cell Components, SSI-19 (Kyoto, Japan), 02.06 - 07.06.2013.
- M. Ender, J. Joos, A. Weber, E. Ivers-Tiffée, The Significance of Tortuosity for the Performance of Lithium-Ion Batteries, 223rd ECS Meeting (Toronto, Canada), 12.05. - 16.05.2013.
- A. Häffelin, J. Joos, M. Ender, A. Weber, E. Ivers-Tiffée, Time-dependent 3D FEM Impedance Model of Mixed-Conducting Solid Oxide Fuel Cell Cathodes, 10th Symposium on Fuel Cell and Battery Modelling and Experimental Validation (Bad Boll, Germany), 19.03. - 20.03.2013.
- J. Joos, M. Ender, T. Carraro, A. Weber, E. Ivers-Tiffée, FIB/SEM Reconstruction and 3D Simulations of SOFC Electrodes, TMS 2013, 142nd Annual Meeting (San Antonio, USA), 03.03 - 07.03.2013.

- E. Ivers-Tiffée, J. Joos, T. Carraro, A. Weber, 3D Analysis and Modeling of Solid Oxide Fuel Cell Cathodes, TMS 2013, 142nd Annual Meeting (San Antonio, USA), 03.03. - 07.03.2013.
- J. Joos, T. Carraro, E. Ivers-Tiffée, 3D reconstruction and modeling of Solid Oxide Fuel Cell electrodes, Meeting of the SOFC Society of Japan, (Tokyo, Japan), 29.10.2012.
- T. Carraro, J. Joos, Numerical Methods for the Simulation and Parameter Estimation of SOFC Microstructures, Meeting of the SOFC Society of Japan, (Tokyo, Japan), 29.10.2012.
- J. Joos, M. Ender, T. Carraro, A. Häffelin, A. Weber, E. Ivers-Tiffée, 3D FEM models for mixed conducting LSCF cathodes, 63rd ISE Meeting (Prague, Czech Republic), 19.08. - 24.08.2012.
- J. Joos, M. Ender, I. Rotscholl, N. Menzler, A. Weber, E. Ivers-Tiffée, Quantification of Ni/YSZ-Anode Microstructure Parameters derived from FIB-tomography, 10th European SOFC Forum (Lucerne, Switzerland), 26.06 - 29.06.2012.
- A. Häffelin, J. Joos, J. Hayd, A. Weber, E. Ivers-Tiffée, 3D FEM Impedance Model for LSCF-Cathodes, 10th European SOFC Forum (Lucerne, Switzerland), 26.06. - 29.06.2012.
- A. Häffelin, J. Joos, M. Ender, E. Ivers-Tiffée, Transient 3D Impedance Model for mixed-conducting SOFC Cathodes, 221th ECS Meeting (Seattle, USA), 2012.
- A. Weber, J. Joos, M. Ender, A. Häffelin, T. Carraro, E. Ivers-Tiffée, Microstructure Reconstruction and Modeling of Multiphase Electrodes, Solid State Electrochemistry for Energy Storage and Conversion 2012 (Heidelberg, Germany), 07.07. - 18.07.2012.
- J. Joos, M. Ender, T. Carraro, A. Weber, E. Ivers-Tiffée, Three Dimensional Analyses and Microstructure Modeling of LSCF Cathodes, 2011MRS Fall Meeting (Boston, USA), 28.11. - 02.12.2011.
- E. Ivers-Tiffée, J. Joos, T. Carraro, M. Ender, FIB/SEM Reconstruction and 3D Simulations of SOFC Electrodes, 2011MRS Fall Meeting (Boston, USA), 28.11. - 02.12.2011.
- A. Häffelin, J. Joos, M. Ender, A. Weber, E. Ivers-Tiffée, Dynamic 3D FEM Model for mixed conducting SOFC Cathodes, MRS Fall Meeting 2011 (Boston, USA), 28.11. - 02.12.2011.
- A. Häffelin, J. Joos, M. Ender, A. Weber, E. Ivers-Tiffée, 3D Model for the Dynamic Simulation of SOFC Cathodes, COMSOL Conference (Stuttgart, Germany), 26.10 – 28.10.2011.

- M. Ender, J. Joos, T. Carraro, E. Ivers-Tiffée, 3D Microstructure Analysis of a Commercial LiFePO4-Cathode, 220th Meeting of The Electrochemical Society (Boston, USA), 09.10. - 14.10.2011.
- J. Joos, M. Ender, T. Carraro, A. Weber, E. Ivers-Tiffée, 3D Microstructure Analysis and Modeling of LSCF cathodes for Solid Oxide Fuel Cells, 62nd Annual ISE Meeting (Niigata, Japan), 11.09. - 16.09.2011.
- J. Joos, M. Ender, T. Carraro, A. Weber, E. Ivers-Tiffée, 3D FEM Model for the Reconstruction of Porous MIEC Cathodes, 18th International Conference on Solid State Ionics (SSI-18) (Warsaw, Poland), 03.07. - 07.07.2011.
- M. Ender, J. Joos, T. Carraro, E. Ivers-Tiffée, FIB/SEM-Tomography For The Exploitation Of Microscale Characteristics Of LiFePO4 - Carbon Black Cathodes, 18th International Conference on Solid State Ionics (SSI-18) (Warsaw, Poland), 03.07. - 07.07.2011.
- J. Joos, T. Carraro, M. Ender, A. Weber, E. Ivers-Tiffée, 3D Finite Element Model for Porous Electrode Microstructures, International Conference on Simulation Technology 2011 (Stuttgart, Germany), 14.06. - 17.06.2011.
- E. Ivers-Tiffée, J. Joos, J. Hayd, High-performance nano-scaled MIEC cathodes for Solid Oxide Fuel Cells, E-MRS Spring Meeting 2011 (Nice, France), 09.05. - 13.05.2011.
- J. Joos, T. Carraro, M. Ender, B. Rüger, A. Weber, E. Ivers-Tiffée, Detailed Microstructure Analysis and 3D Simulations of Porous Electrodes, 219th Meeting of The Electrochemical Society (Montreal, Canada), 01.05. - 06.05.2011.
- M. Ender, J. Joos, T. Carraro, E. Ivers-Tiffée, 3D Microstructure Characterization of a Composite Cathode for Lithium-Ion Cells, 219th Meeting of The Electrochemical Society (Montreal, Canada), 01.05. - 06.05.2011.
- A. Weber, A. Utz, J. Joos, E. Ivers-Tiffée, H. Störmer, D. Gerthsen, V. Yurkiv, H.-R. Volpp, W.G. Bessler, Electrooxidation of Reformate Gases at Model Anodes, SOFC XII (Montreal, Canada), 01.05. - 06.05.2011.
- A. Utz, J. Joos, A. Weber, E. Ivers-Tiffée, Electrochemical Oxidation at SOFC Anodes: Comparison of Patterned Ni Anodes and Technical Ni/8YSZ Cermet Anodes, 219th Meeting of The Electrochemical Society (Montreal, Canada), 01.05. - 06.05.2011.
- J. Joos, T. Carraro, M. Ender, A. Weber, E. Ivers-Tiffée, 3D Rekonstruktion und Modellierung einer porösen Elektrode für die Hochtemperaturbrennstoffzelle SOFC, DKG-Jahrestagung 2011 (Saarbrücken, Deutschland), 28.03. - 30.03.2011.
- T. Carraro, J. Joos, A. Weber, E. Ivers-Tiffée, 3D FEM-model for the Reconstruction of a Porous SOFC Cathode, MRS Fall Meeting 2010 (Boston, USA), 29.11. - 02.12.2010.

- J. Joos, B. Rüger, A. Weber, T. Carraro, E. Ivers-Tiffée, 3D Reconstruction of Porous Electrodes and Microstructure Modelling, First International Conference on Materials for Energy (Karlsruhe, Germany), 04.07. - 08.07.2010.
- J. Joos, T. Carraro, A. Weber, E. Ivers-Tiffée, Reconstruction of Porous Electrodes by FIB/SEM for Detailed Microstructure Modeling, 9th European SOFC Forum (Lucerne, Switzerland), 29.06. - 02.07.2010.
- J. Joos, T. Carraro, B. Rüger, A. Weber, E. Ivers-Tiffée, Electrode Reconstruction by FIB/SEM and Microstructure Modeling, 217th Meeting of The Electrochemical Society (Vancouver, Canada), 25.04. - 30.04.2010.
- J. Joos, B. Rüger, T. Carraro, A. Weber, E. Ivers-Tiffée, 3D Reconstruction and Modelling of Porous Electrodes, International Conference on Electroceramics (ICE-2009) (New Delhi, India), 13.12. - 17.12.2009.
- B. Rüger, J. Joos, T. Carraro, A. Weber, E. Ivers-Tiffée, 3D Electrode Microstructure Reconstruction and Modelling, SOFC XI (Vienna, Austria), 04.10. - 09.10.2009.
- B. Rüger, T. Carraro, J. Joos, V. Heuveline, E. Ivers-Tiffée, A Numerical Method for the Optimization of Electrode Microstructure, 6th Symposium on Fuel Cell Modelling and Experimental Validation (Bad Herrenalb, Germany), 25.03. - 26.03.2009.
- B. Rüger, J. Joos, A. Weber, E. Ivers-Tiffée, Microstructure Modelling and Performance Evaluation of Mixed Conducting Cathodes for SOFC Application, MRS Fall Meeting 2008 (Boston MA, USA), 01.12. - 05.12.2008.
- T. Carraro, B. Rüger, J. Joos, V. Heuveline, A detailed FE model for the simulation of MIEC cathode microstructure of a SOFC, Conference on Modeling, Simulation and Optimization of Complex Processes (MOSOCOP 2008) (Heidelberg, Germany), 21.07. - 25.07.2008.
- B. Rüger, T. Carraro, J. Joos, E. Ivers-Tiffée, A Numerical Method for the Optimization of MIEC Cathode Microstructures, 8th EUROPEAN SOFC FORUM (Lucerne, Switzerland), 30.06. - 04.07.2008.
- B. Rüger, T. Carraro, J. Joos, A. Weber, E. Ivers-Tiffée, Modeling the microstructure of electrodes, 2nd Mini Symposium on Solid Oxide Fuel Cells (Karlsruhe, Germany), 27.06. - 27.06.2008.

G. Symbols

Symbol	Description	Unit/Value
A	area	nm^2
a_i	surface ares density (of phase i)	µm
ASR	specific resistance	Ωm^2
ASR_{cat}	area specific resistance of the cathode	Ωm^2
c_{mc}	oxygen lattice sites	mol/m^3
c_{O_2}	oxygen ion concentration	mol/m^3
$c_{O^{2-},eq}$	oxygen ion concentration at equilibrium	mol/m^3
D^δ	chemical diffusion coefficient	m^2/s
$D_{O_2N_2}$	binary diffusion coefficient	m^2/s
$D^k_{O_2}(d_{pore})$	Knudsen diffusion coefficient	m^2/s
d_p	particle- or pore size (diameter)	m
F	Faraday constant	$9.649 \cdot 10^4$ C/mol
ΔG	change in Gibbs free energy	J/mol
$j_{0,el}$	exchange current density of electrode	A/m^2
J_{O_2}	flux of oxygen molecules	$mol/(sm^2)$
k^δ	chemical surface exchange coefficient	m/s
L	geometric length (e.g. electrode thickness)	m
L_{eff}	effective (average) path length	m
l_{TPB}	triple-phase boundary density	μm^{-2}
l_δ	penetration depth (characteristic length)	m
l_v	edge length of the cubes/voxels	m
LSR_{ct}	line-specific resistance	
M_i	molar mass of the gas component i	kg/mol
N	(total) number of voxel	-
N_i	number of voxel of phase i	-
n_e	number of electrons	-
pO_2	oxygen partial pressure	atm
R	ideal gas constant	$8.314\,JK^{-1}\,mol^{-1}$
R_{chem}	characteristic resistance	Ω
R_{pol}	electrode polarization resistance	Ωcm^2
T	temperature	K
t_{chem}	characteristic time constant	s
t_{sint}	sintering time	h
T_{sint}	sintering temperature	K
U	Voltage	V
U_{cell}	cell voltage	V
U_N	Nernst voltage	V

Symbol	Description	Unit/Value
U_{th}	theoretical Nernst voltage	V
V_i	volume fraction of phase *i*	%
η_{ohm}	ohmic overpotential	V
σ_{ion}	ionic conductivity	S/m
x_{O_2}	oxygen molar fraction in the pores	-
γ_o	thermodynamic factor for oxygen ion concentration	-
γ_v	thermodynamic factor for vacancies	-
δ	oxygen nonstoichiometry	-
τ_i	tortuosity of phase i	-
ε	porosity fraction	%
η_{act}	activation overpotential	V
$\eta_{cathode}$	total cathode overpotential	V
η_{Model}	voltage losses (or overpotential) of the model	V
$\tilde{\mu}_i$	electrochemical potential of *i*	J/mol
σ_i	conductivity of *i*	S/m
Φ_i	potentials of layer *i*	V

Indices

an	anode
CC	current collector (gas channel)
CE	counter electrode
el	electrode (anode or cathode)
elyt	electrolyte
cat	cathode
sint	sintering
vf	voxel-face

H. Acronyms

AFL	Anode Functional Layer
ASC	Anode Supported Cell
ASR	Area Specific Resistance
BOP	Balance of Plant
BSE	Back-Scattered Electrons
BSCF	Barium Strontium Cobalt Ferrite
CM	Centroid Method
ETD	Euclidean Distance Transformation
EDX	Energy Dispersive X-ray
EIS	Electrochemical Impedance Spectroscopy

EMF	Electromotive Force
ETD	Everhart-Thornley Detector
FEM	Finite Element Method
FIB	Focused Ion Beam
FVM	Finite Volume Method
FZJ	Forschungszentrum Jülich
GDC	Gadolinium-doped Ceria
HPC	High Performance Computing
IAM-WET	Institute for Applied Materials
IL	In-Lense Detector
LBM	Lattice Boltzmann Method
LSC	Lanthanum Strontium Cobalt
LSCF	Lanthanum Strontium Cobalt Ferrite
LSM	Lanthanum Strontium Manganite
MC	Marching Cubes (Method)
MEA	Membrane Electrode Assembly
MIEC	Mixed Ionic/Electronic Conducting
NiO	Nickel-Oxide
OCV	Open Circuit Voltage
RAM	Random-access memory
ROI	Region of Interest
RVE	Representative Volume Element
SE	Secondary Electrons
SA	Surface Area
SCC	Steinbuch Centre of Computing
SEM	Scanning Electron Microscopy
SERVE	Stochastic Equivalent Representative Volume Element
SOFC	Solid Oxide Fuel Cell
TPB	Triple-Phase Boundary
VM	Voxel-Mesh
YSZ	Yttria Stabilized Zirconia

Bibliography

[1] S. C. Singhal. *High-temperature Solid Oxide Fuel Cells: Fundamentals, Design and Applications: Fundamentals, Design and Applications*. Elsevier, 2003.

[2] A. Leonide. *SOFC Modelling and Parameter Identification by Means of Impedance Spectroscopy*. PhD thesis, Karlsruher Institut für Technologie (KIT), Karlsruhe, 2010.

[3] J. Hayd. *Nanoskalige Kathoden für den Einsatz in Festelektrolyt-Brennstoffzellen bei abgesenkten Betriebstemperaturen*. PhD thesis, Karlsruher Institut für Technologie (KIT), 2012.

[4] P. Holtappels and U. Stimming. *Solid oxide fuel cells (SOFC)*. John Wiley and Sons, Ltd, 2010.

[5] M. J. Heneka. *Alterung der Festelektrolyt-Brennstoffzelle unter thermischen und elektrischen Lastwechseln*. PhD thesis, Universitlät Karlsruhe, 2006.

[6] W. Nernst. Die elektromotorische Wirksamkeit der Jonen. *Zeitschrift für Physikalische Chemie*, 4(2):129–181, 1889.

[7] E. Ivers-Tiffée and A. V. Virkar. Electrode Polarisation. In S.C. Singhal and K. Kendall, editors, *High Temperature Solid Oxide Fuel Cells - Fundamentals, Design and Applications*, pages 231–260. Oxford: Elsevier Ltd, 2003.

[8] A. Weber. *Entwicklung von Kathodenstrukturen für die Hochtemperatur-Brennstoffzelle SOFC*. PhD thesis, Universität Karlsruhe, 2003.

[9] A. Weber and E. Ivers-Tiffée. Materials and concepts for solid oxide fuel cells (SOFCs) in stationary and mobile applications. *Journal of Power Sources*, 127(1):273–283, 2004.

[10] A. C. Müller, J. R. Opfermann, and E. Ivers-Tiffée. Modelling and optimisation of solid electrolyte sintering behaviour by thermokinetic analysis. *Thermochimica acta*, 414(1):11–17, 2004.

[11] M. E. Orazem and B. Tribolett. *Electrochemical Impedance Spectroscopy*. John Wiley and Sons, New York, 2008.

[12] P. Shearing. *Characterisation of Solid Oxide Fuel Cell Electrode Microstructures in Three Dimensions*. PhD thesis, Imperial Colleague, 2009.

[13] S. Primdahl and M. Mogensen. Gas diffusion impedance in characterization of solid oxide fuel cell anodes. *Journal of the Electrochemical Society*, 146(8):2827–2833, 1999.

[14] J. W. Kim, A. V. Virkar, K. Z. Fung, K. Mehta, and S. C. Singhal. Polarization Effects in Intermediate Temperature, Anode-Supported Solid Oxide Fuel Cells. *Journal of The Electrochemical Society*, 146(1):69–78, 1999.

[15] N. Q. Minh. Ceramic Fuel Cells. *Journal of the American Ceramic Society*, 76(3):563–588, 1993.

[16] O. Yamamoto. Solid oxide fuel cells: Fundamental aspects and prospects. *Electrochimica Acta*, 45(15-16):2423–2435, 2000.

[17] F. Tietz, H.-P. Buchkremer, and D. Stöver. Components manufacturing for solid oxide fuel cells. *Solid State Ionics*, 152-153(0):373–381, 2002.

[18] N. P. Brandon, S. Skinner, and B. C. H. Steele. Recent advances in materials for fuel cells. *Annual Review of Materials Research*, 33(1):183–213, 2003.

[19] S. M. Haile. Fuel cell materials and components. *Acta Materialia*, 51(19):5981–6000, 2003.

[20] S. P. S. Badwal. Zirconia-based solid electrolytes: microstructure, stability and ionic conductivity. *Solid State Ionics*, 52(1):23–32, 1992.

[21] E. Ivers-Tiffée and W. von Münch. *Werkstoffe der Elektrotechnik*. Springer-Verlag, 2007.

[22] B. Dalslet, P. Blennow, P. V. Hendriksen, N. Bonanos, D. Lybye, and M. Mogensen. Assessment of doped ceria as electrolyte. *Journal of Solid State Electrochemistry*, 10(8):547–561, 2006.

[23] K. Eguchi, T. Hatagishi, and H. Arai. Power generation and steam electrolysis characteristics of an electrochemical cell with a zirconia-or ceria-based electrolyte. *Solid State Ionics*, 86:1245–1249, 1996.

[24] F. Tietz, A. Mai, and D. Stöver. From powder properties to fuel cell performance - A holistic approach for SOFC cathode development. *Solid State Ionics*, 179(27-32):1509–1515, 2008.

[25] J. Richter, P. Holtappels, T. Graule, T. Nakamura, and L. Gauckler. Materials design for perovskite SOFC cathodes. *Monatshefte für Chemie*, 140:985–999, 2012.

[26] B. Rüger. *Mikrostrukturmodellierung von Elektroden für die Festelektrolytbrennstoffzelle*. PhD thesis, Universität Karlsruhe, 2009.

[27] C. Endler-Schuck. *Alterungsverhalten mischleitender LSCF Kathoden für Hochtemperatur-Festoxid-Brennstoffzellen (SOFCs)*. PhD thesis, Karlsruher Institut für Technologie (KIT), 2012.

[28] O. Yamamoto, Y. Takeda, R. Kanno, and M. Noda. Perovskite-type oxides as oxygen electrodes for high temperature oxide fuel cells. *Solid State Ionics*, 22(2-3):241–246, 1987.

[29] S. Uhlenbruck, N. Jordan, D. Sebold, H. P. Buchkremer, V. A. C. Haanappel, and D. Stöver. Thin film coating technologies of (Ce,Gd)O2-d interlayers for application in ceramic high-temperature fuel cells. *Thin Solid Films*, 515(7-8):4053–4060, 2007.

[30] A. Mai, V. A. C. Haanappel, S. Uhlenbruck, F. Tietz, and D. Stöver. Ferrite-based perovskites as cathode materials for anode-supported solid oxide fuel cells: Part I. Variation of composition. *Solid State Ionics*, 176(15):1341–1350, 2005.

[31] A. Mai, V. A. C. Haanappel, F. Tietz, and D. Stöver. Ferrite-based perovskites as cathode materials for anode-supported solid oxide fuel cells: part II. Influence of the CGO interlayer. *Solid State Ionics*, 177(19):2103–2107, 2006.

[32] E. Ivers-Tiffée, A. Weber, and H. Schichlein. O_2-reduction at high temperatures: SOFC. In *Handbook of Fuel Cells*, volume 2, pages 587–600. Wiley Online Library, 2003.

[33] J. Mizusaki, Y. Yonemura, H. Kamata, K. Ohyama, N. Mori, H. Takai, H. Tagawa, M. Dokiya, K. Naraya, T. Sasamoto, et al. Electronic conductivity, seebeck coefficient, defect and electronic structure of nonstoichiometric La_{1-x} Sr_x MnO_3. *Solid State Ionics*, 132(3):167–180, 2000.

[34] S. P. Jiang. Development of lanthanum strontium manganite perovskite cathode materials of solid oxide fuel cells: a review. *Journal of Materials Science*, 43(21):6799–6833, 2008.

[35] F. Tietz, V. A. C Haanappel, A. Mai, J. Mertens, and D. Stöver. Performance of lscf cathodes in cell tests. *Journal of Power Sources*, 156(1):20–22, 2006.

[36] A. Mai, M. Becker, W. Assenmacher, F. Tietz, D. Hathiramani, E. Ivers-Tiffée, D. Stöver, and W. Mader. Time-dependent performance of mixed-conducting SOFC cathodes. *Solid State Ionics*, 177(19):1965–1968, 2006.

[37] A. Leonide, V. Sonn, A. Weber, and E. Ivers-Tiffée. Evaluation and modeling of the cell resistance in anode-supported solid oxide fuel cells. *Journal of The Electrochemical Society*, 155(1):B36–B41, 2008.

[38] C. Endler, A. Leonide, A. Weber, F. Tietz, and E. Ivers-Tiffée. Time-dependent electrode performance changes in intermediate temperature solid oxide fuel cells. *Journal of The Electrochemical Society*, 157(2):B292–B298, 2010.

[39] C. Endler-Schuck, J. Joos, C. Niedrig, A. Weber, and E. Ivers-Tiffée. The Chemical Oxygen Surface Exchange and Bulk Diffusion Coefficient Determined by Impedance Spectroscopy of Porous $La_{0.58}\,Sr_{0.4}\,Co_{0.2}\,Fe_{0.8}\,O_{3-\delta}$ (LSCF) Cathodes. *Solid State Ionics*, 269:67–79, 2015.

[40] M. Mori, T. Yamamoto, H. Itoh, H. Inaba, and H. Tagawa. Thermal expansion of nickel-zirconia anodes in solid oxide fuel cells during fabrication and operation. *Journal of the Electrochemical Society*, 145(4):1374–1381, 1998.

[41] E. Ivers-Tiffée, W. Wersing, M. Schiessl, and H. Greiner. Ceramic and metallic components for a planar sofc. *Berichte der Bunsengesellschaft für physikalische Chemie*, 94(9):978–981, 1990.

[42] F. Tietz, F. J Dias, D. Simwonis, and D. Stöver. Evaluation of commercial nickel oxide powders for components in solid oxide fuel cells. *Journal of the European Ceramic Society*, 20(8):1023–1034, 2000.

[43] D. Fouquet, A. C. Müller, A. Weber, and E. Ivers-Tiffée. Kinetics of oxidation and reduction of Ni/YSZ cermets. *Ionics*, 9:103–108, 2003.

[44] D. Sarantaridis and A. Atkinson. Redox Cycling of Ni-Based Solid Oxide Fuel Cell Anodes: A Review. *Fuel Cells*, 7(3):246–258, 2007.

[45] A. Kromp. *Model-based Interpretation of the Performance and Degradation of Reformate Fueled Solid Oxide Fuel Cells*. PhD thesis, Karlsruher Institut für Technologie (KIT), Karlsruhe, 2013. Zugl.: Karlsruhe, KIT, Diss., 2013.

[46] A. Atkinson, S. A. Barnett, R. J. Gorte, J. T. S. Irvine, A. J. McEvoy, M. Mogensen, S. C. Singhal, and J. Vohs. Advanced anodes for high-temperature fuel cells. *Nature Materials*, Jan; 3:17–27, 2004.

[47] W. Z. Zhu and S. C. Deevi. A review on the status of anode materials for solid oxide fuel cells. *Materials Science and Engineering: A*, 362(1-2):228–239, 2003.

[48] S. P. Jiang and S. H. Chan. A review of anode materials development in solid oxide fuel cells. *Journal of Materials Science*, 39:4405–4439, 2004.

[49] E. Fontell, T. Kivisaari, N. Christiansen, J.-B. Hansen, and J. Pålsson. Conceptual study of a 250kW planar SOFC system for CHP application. *Journal of Power Sources*, 131(1):49–56, 2004.

[50] D. Waldbillig, A. Wood, and D. G. Ivey. Electrochemical and microstructural characterization of the redox tolerance of solid oxide fuel cell anodes. *Journal of Power Sources*, 145(2):206–215, 2005.

[51] M. Kleitz and F. Petitbon. Optimized sofc electrode microstructure. *Solid State Ionics*, 92(1):65–74, 1996.

[52] S. B. Adler. Factors governing oxygen reduction in solid oxide fuel cell cathodes. *Chemical reviews*, 104(10):4791–4844, 2004.

[53] D. Herbstritt. *Entwicklung und Modellierung einer leistungsfähigen Kathodenstruktur für die Hochtemperaturbrennstoffzelle SOFC*. PhD thesis, Universität Karlsruhe, 2003.

[54] T. Horita, K. Yamaji, N. Sakai, H. Yokokawa, T. Kawada, and T. Kato. Oxygen reduction sites and diffusion paths at $La_{0.9}$ $Sr_{0.1}$ MnO_{3-x}/yttria-stabilized zirconia interface for different cathodic overvoltages by secondary-ion mass spectrometry. *Solid State Ionics*, 127(1):55–65, 2000.

[55] T. Horita, K. Yamaji, N. Sakai, Y. Xiong, T. Kato, H. Yokokawa, and T. Kawada. Imaging of oxygen transport at sofc cathode/electrolyte interfaces by a novel technique. *Journal of Power Sources*, 106(1):224–230, 2002.

[56] J. Fleig, H.-R. Kim, J. Jamnik, and J. Maier. Oxygen Reduction Kinetics of Lanthanum Manganite (LSM) Model Cathodes: Partial Pressure Dependence and Rate-Limiting Steps. *Fuel Cells*, 8(5):330–337, 2008.

[57] R. Radhakrishnan, A. V. Virkar, and S. C. Singhal. Estimation of charge-transfer resistivity of $La_{0.8}Sr_{0.2}Mno_3$ cathode on$Y_{0.16}$ $Zr_{0.84}o_2$ y 0.16 zr0. 84 o 2 electrolyte using patterned electrodes. *Journal of The Electrochemical Society*, 152(1):A210–A218, 2005.

[58] B. Yildiz, S. McEuen, Y. Shao-Horn, et al. Probing Oxygen Reduction Reaction Kinetics of Sr-Doped $LaMnO_3$ Supported on Y_2O_3-Stabilized ZrO_2 EIS of Dense, Thin-Film Microelectrodes. *Journal of The Electrochemical Society*, 154(4):B427–B438, 2007.

[59] R. Merkle, J. Maier, and J. Fleig. Mechanistic understanding and electrochemical modeling of mixed conducting (SOFC) electrodes. In *Handbook of Fuel Cells*, volume 5, pages 425–440. Wiley Online Library, 2003.

[60] S. B. Adler, J. A. Lane, and B. C. H. Steele. Electrode Kinetics of Porous Mixed-Conducting Oxygen Electrodes. *Journal of The Electrochemical Society*, 143(11):3554–3564, 1996.

[61] Roger Armand De Souza. *Ionic transport in acceptor-doped perovskites*. PhD thesis, Imperial College London (University of London), 1996.

[62] S. B. Adler and Bessler. W. G. Elementary kinetic modeling of solid oxide fuel cell electrode reactions. In *Handbook of Fuel Cells*, volume 5, pages 441–462. Wiley Online Library, 2011.

[63] J. Maier. *Physical chemistry of ionic materials: ions and electrons in solids*. John Wiley & Sons, 2004.

[64] J. Maier. On the correlation of macroscopic and microscopic rate constants in solid state chemistry. *Solid State Ionics*, 112(3):197–228, 1998.

[65] H. J. M. Bouwmeester and A. J. Burggraaf. Dense ceramic membranes for oxygen separation. In *Membrane Science and Technology*, volume 4, chapter 10, pages 435–528. Elsevier, 1996.

[66] J. A. Lane, S. J. Benson, D. Waller, and J. A. Kilner. Oxygen transport in $La_{0.6}Sr_{0.4}Co_{0.2}Fe_{0.8}o_{3-\delta}$. *Solid State Ionics*, 121(1):201–208, 1999.

[67] S. Wang, A. Verma, Y. L. Yang, A. J. Jacobson, and B. Abeles. The effect of the magnitude of the oxygen partial pressure change in electrical conductivity relaxation measurements: oxygen transport kinetics in $La_{0.5}Sr_{0.5}CoO_{3-\delta}$. *Solid State Ionics*, 140(1):125–133, 2001.

[68] N. Hildenbrand, B. A. Boukamp, P. Nammensma, and D. H. A. Blank. Improved cathode/electrolyte interface of sofc. *Solid state ionics*, 192(1):12–15, 2011.

[69] Y. Lu, C. Kreller, and S. B. Adler. Measurement and Modeling of the Impedance Characteristics of Porous $La_{1-x}Sr_xCoO_{3-\delta}$ Electrodes. *Journal of The Electrochemical Society*, 156(4):B513–B525, 2009.

[70] R. Sayers, M. Rieu, P. Lenormand, F. Ansart, J. A. Kilner, and S. J. Skinner. Development of lanthanum nickelate as a cathode for use in intermediate temperature solid oxide fuel cells. *Solid State Ionics*, 192(1):531–534, 2011.

[71] N. Hildenbrand, P. Nammensma, D. H. A. Blank, H. J. M. Bouwmeester, and B. A. Boukamp. Influence of configuration and microstructure on performance of $La_2NiO_{4+\delta}$ intermediate-temperature solid oxide fuel cells cathodes. *Journal of Power Sources*, 238:442–453, 2013.

[72] O. A. Marina, C. Bagger, S. Primdahl, and M. Mogensen. A solid oxide fuel cell with a gadolinia-doped ceria anode: preparation and performance. *Solid State Ionics*, 123(1):199–208, 1999.

[73] H. Uchida, S. Suzuki, and M. Watanabe. High performance electrode for medium-temperature solid oxide fuel cells mixed conducting ceria-based anode with highly dispersed ni electrocatalysts. *Electrochemical and solid-state letters*, 6(9):A174–A177, 2003.

[74] S. Tao and J. T. S. Irvine. Optimization of mixed conducting properties of Y_2O_3–ZrO_2–TiO_2 and Sc_2O_3–Y_2 O_3–ZrO_2–TiO_2 solid solutions as potential sofc anode materials. *Journal of solid state chemistry*, 165(1):12–18, 2002.

[75] A. Utz, A. Leonide, A. Weber, and E. Ivers-Tiffée. Studying the co–co2 characteristics of sofc anodes by means of patterned ni anodes. *Journal of Power Sources*, 196(17):7217–7224, 2011.

[76] W. G. Bessler, M. Vogler, H. Störmer, D. Gerthsen, A. Utz, A. Weber, and E. Ivers-Tiffée. Model anodes and anode models for understanding the mechanism of hydrogen oxidation in solid oxide fuel cells. *Physical Chemistry Chemical Physics*, 12(42):13888–13903, 2010.

[77] A. Bieberle and L. J. Gauckler. Reaction mechanism of Ni pattern anodes for solid oxide fuel cells. *Solid State Ionics*, 135(1):337–345, 2000.

[78] A. Bieberle, L. P. Meier, and L. J. Gauckler. The electrochemistry of Ni pattern anodes used as solid oxide fuel cell model electrodes. *Journal of the electrochemical society*, 148(6):A646–A656, 2001.

[79] J. Mizusaki, H. Tagawa, T. Saito, K.i Kamitani, T. Yamamura, K. Hirano, S. Ehara, T. Takagi, T. Hikita, M. Ippommatsu, et al. Preparation of Nickel Pattern Electrodes on YSZ and Their Electrochemical Properties in H_2–H_2O Atmospheres. *Journal of the Electrochemical Society*, 141(8):2129–2134, 1994.

[80] C. W. Tanner, K.-Z. Fung, and A. V. Virkar. The effect of porous composite electrode structure on solid oxide fuel cell performance: I. theoretical analysis. *Journal of The Electrochemical Society*, 144(1):21–30, 1997.

[81] A. Utz, J. Joos, A. Weber, and E. Ivers-Tiffée. Electrochemical oxidation at sofc anodes: Comparison of patterned nickel anodes and nickel/8ysz cermet anodes. *ECS Transactions*, 35(1):1669–1682, 2011.

[82] A. Utz, H. Störmer, D. Gerthsen, A. Weber, and E. Ivers-Tiffée. Microstructure stability studies of ni patterned anodes for sofc. *Solid State Ionics*, 192(1):565–570, 2011.

[83] A. P. Cocco, G. J. Nelson, W. M. Harris, A. Nakajo, T. D. Myles, A. M. Kiss, J. J. Lombardo, and W. K. S. Chiu. Three-dimensional microstructural imaging methods for energy materials. *Physical Chemistry Chemical Physics*, 15(39):16377–16407, 2013.

[84] P. R. Shearing, D. J. L. Brett, and N. P. Brandon. Towards intelligent engineering of sofc electrodes: a review of advanced microstructural characterisation techniques. *International Materials Reviews*, 55(6):347–363, 2010.

[85] J. Laurencin, R. Quey, G. Delette, H. Suhonen, P. Cloetens, and P. Bleuet. Characterisation of solid oxide fuel cell ni–8ysz substrate by synchrotron x-ray nanotomography: from 3d reconstruction to microstructure quantification. *Journal of Power Sources*, 198:182–189, 2012.

[86] J. R. Wilson, W. Kobsiriphat, R. Mendoza, H.-Y. Chen, J. M. Hiller, D. J. Miller, K. Thornton, P. W. Voorhees, S. B. Adler, and S. A. Barnett. Three-dimensional reconstruction of a solid-oxide fuel-cell anode. *Nature materials*, 5(7):541–544, 2006.

[87] L. Reimer and G. Pfefferkorn. *Raster-Elektronenmikroskopie*. Springer, 1973.

[88] V. Mittal and N. B. Matsko. *Analytical Imaging Techniques for Soft Matter Characterization*. Springer, 2012.

[89] J. Ackermann. Manual for the SUPRA (VP) and ULTRA Scanning Electron Microscopes. Manual, Carl Zeiss NTS GmbH, 2005.

[90] J. I. Goldstein, editor. *Scanning electron microscopy and x-ray microanalysis*. Kluwer Academic/Plenum Publishers, New York [u.a.], 3. ed. edition, 2003.

[91] P. R. Munroe. The application of focused ion beam microscopy in the material sciences. *Materials characterization*, 60(1):2–13, 2009.

[92] J. Orloff, L. Swanson, and M. Utlaut. *High Resolution Focused Ion Beams: FIB and its Applications: Fib and Its Applications: The Physics of Liquid Metal Ion Sources and Ion Optics and Their Application to Focused Ion Beam Technology*. Springer, 2003.

[93] L. A. Giannuzzi and F. A. Stevie. *Introduction to focused ion beams: instrumentation, theory, techniques and practice*. Springer, 2005.

[94] W. Zinth and U. Zinth. *Optik: Lichtstrahlen-Wellen-Photonen*. Oldenbourg Verlag, 2013.

[95] M. Von Ardenne. Das elektronen-rastermikroskop. *Zeitschrift für Physik*, 109(9-10):553–572, 1938.

[96] K. Thyden, Y.-L. Liu, and J. B. Bilde-Sørensen. Microstructural characterization of sofc ni–ysz anode composites by low-voltage scanning electron microscopy. *Solid State Ionics*, 178(39):1984–1989, 2008.

[97] H. Iwai, N. Shikazono, T. Matsui, H. Teshima, M. Kishimoto, R. Kishida, D. Hayashi, K. Matsuzaki, D. Kanno, M. Saito, et al. Quantification of sofc anode microstructure based on dual beam fib-sem technique. *Journal of Power Sources*, 195(4):955–961, 2010.

[98] Moses Johannes Ender. *Mikrostrukturelle Charakterisierung, Modellentwicklung und Simulation poroeser Elektroden fuer Lithiumionenzellen*. PhD thesis, Karlsruher Institut für Technologie (KIT), Karlsruhe, 2014. Zugl.: Karlsruhe, KIT, Diss., 2014.

[99] J. Ohser and K. Schladitz. *3D images of materials structures: processing and analysis*. John Wiley & Sons, 2009.

[100] R. C. González, R. E. Woods, and S. L. Eddins. *Digital image processing using MATLAB*. Gatesmark Pub., 2009.

[101] B. Rüger, J. Joos, A. Weber, T. Carraro, and E. Ivers-Tiffée. 3D electrode microstructure reconstruction and modelling. *ECS Transactions*, 25(2):1211–1220, 2009.

[102] P. R. Shearing, J. Golbert, R. J. Chater, and N. P. Brandon. 3d reconstruction of sofc anodes using a focused ion beam lift-out technique. *Chemical Engineering Science*, 64(17):3928–3933, 2009.

[103] J. R. Wilson, M. Gameiro, K. Mischaikow, W. Kalies, P. W. Voorhees, and S. A. Barnett. Three-dimensional analysis of solid oxide fuel cell Ni-YSZ anode interconnectivity. *Microscopy and Microanalysis*, 15(01):71–77, 2009.

[104] J. Joos, T. Carraro, M. Ender, B. Rüger, A. Weber, and E. Ivers-Tiffée. Detailed microstructure analysis and 3d simulations of porous electrodes. *ECS Transactions*, 35(1):2357–2368, 2011.

[105] O. Demirkaya, M. H. Asyali, and P. K. Sahoo. *Image processing with MATLAB: applications in medicine and biology*. CRC Press, 2011.

[106] P. Perona and J. Malik. Scale-space and edge detection using anisotropic diffusion. *Pattern Analysis and Machine Intelligence, IEEE Transactions on*, 12(7):629–639, 1990.

[107] J. Weickert. *Anisotropic diffusion in image processing*, volume 1. Teubner Stuttgart, 1998.

[108] D. Klotz. *Characterization and Modeling of Electrochemical Energy Conversion Systems by Impedance Techniques*. Schriften des Instituts für Werkstoffe der Elektrotechnik, Karlsruher Institut für Technologie ; 23. KIT Scientific Publishing, 2012.

[109] A. Leonide, S. Ngo Dinh, A. Weber, and E. Ivers-Tiffée. Performance limiting factors in anode supported sofc. *Proceedings of the 8th European Solid Oxide Fuel Cell Forum*, page A0501, 2008.

[110] D. Simwonis, H. Thülen, F. J. Dias, A. Naoumidis, and D. Stöver. Properties of ni/ysz porous cermets for sofc anode substrates prepared by tape casting and coat-mix® process. *Journal of Materials Processing Technology*, 92:107–111, 1999.

[111] R. Mücke, N. H. Menzler, H. P. Buchkremer, and D. Stöver. Cofiring of thin zirconia films during sofc manufacturing. *Journal of the American Ceramic Society*, 92(s1):S95–S102, 2009.

[112] H.P. Buchkremer, U. Diekmann, and D. Stöver. Components Manufacturing and Stack Integration of an Anode Supported Planar SOFC System. *Proceedings of the 2nd European Solid Oxide Fuel Cell Forum*, 1:221–228, 1996.

[113] D. Stöver, H. P. Buchkremer, and J. P. P. Huijsmans. MEA/cell preparation methods: Europe/USA. In W. Vielstich, H. A. Gasteiger, and A. Lamm, editors, *Handbook of Fuel Cells — Fuel Cell Technology and Applications Vol. 4*, pages 1015–1029. John Wiley & Sons, Ltd, 2003.

[114] N. H. Menzler, W. Schafbauer, F. Han, O. Büchler, R. Mücke, H. P. Buchkremer, and D. Stöver. Development of High Power Density Solid Oxide Fuel Cells (SOFCs) for Long-Term Operation. *Materials Science Forum*, 654-656:2875–2878, 2010.

[115] E. Ivers-Tiffée, J. Hayd, D. Klotz, A. Leonide, F. Han, and A. Weber. Performance analysis and development strategies for solid oxide fuel cells. *ECS Transactions*, 35(1):1965–1973, 2011.

[116] G. K. Steinke. *Elektrochemische Charakterisierung mikrostrukturell modifizierter Kathoden für die SOFC*. Study project, Karlsruher Institut für Technologie (KIT), 2011.

[117] J. Joos, T. Carraro, A. Weber, and E. Ivers-Tiffée. Reconstruction of porous electrodes by fib/sem for detailed microstructure modeling. *Journal of Power Sources*, 196(17):7302–7307, 2011.

[118] J. Joos, M. Ender, T. Carraro, A. Weber, and E. Ivers-Tiffée. Representative volume element size for accurate solid oxide fuel cell cathode reconstructions from focused ion beam tomography data. *Electrochimica Acta*, 82:268–276, 2012.

[119] T. Carraro, J. Joos, B. Rüger, A. Weber, and E. Ivers-Tiffée. 3D finite element model for reconstructed mixed-conducting cathodes: I. Performance quantification. *Electrochimica Acta*, 77:315–323, 2012.

[120] J. Joos, B. Rüger, T. Carraro, A. Weber, and E. Ivers-Tiffée. Electrode reconstruction by fib/sem and microstructure modeling. *ECS Transactions*, 28(11):81–91, 2010.

[121] J. Joos, M. Ender, I. Rotscholl, N. H. Menzler, and E. Ivers-Tiffée. Quantification of double-layer ni/ysz fuel cell anodes from focused ion beam tomography data. *Journal of Power Sources*, 246:819–830, 2014.

[122] L. Holzer and M. Cantoni. Review of FIB-tomography. In *Nanofabrication Using Focused Ion and Electron Beams: Principles and Applications*, volume 559201222, chapter 11, pages 410–435. Oxford University Press, 2011.

[123] J. R. Wilson, A. T. Duong, M. Gameiro, H.-Y. Chen, K. Thornton, D. R. Mumm, and S. A. Barnett. Quantitative three-dimensional microstructure of a solid oxide fuel cell cathode. *Electrochemistry Communications*, 11(5):1052–1056, 2009.

[124] K. Yakal-Kremski, J. S. Cronin, Y.-CK Chen-Wiegart, J. Wang, and S. A. Barnett. Studies of solid oxide fuel cell electrode evolution using 3D tomography. *Fuel Cells*, 13(4):449–454, 2013.

[125] J. S. Cronin, J. R. Wilson, and S. A. Barnett. Impact of pore microstructure evolution on polarization resistance of Ni-Yttria-stabilized zirconia fuel cell anodes. *Journal of Power Sources*, 196(5):2640–2643, 2011.

[126] N. Shikazono, D. Kanno, K. Matsuzaki, H. Teshima, S. Sumino, and N. Kasagi. Numerical assessment of sofc anode polarization based on three-dimensional model microstructure reconstructed from fib-sem images. *Journal of The Electrochemical Society*, 157(5):B665–B672, 2010.

[127] D. Kanno, N. Shikazono, N. Takagi, K. Matsuzaki, and N. Kasagi. Evaluation of sofc anode polarization simulation using three-dimensional microstructures reconstructed by fib tomography. *Electrochimica Acta*, 56(11):4015–4021, 2011.

[128] K. Matsuzaki, N. Shikazono, and N. Kasagi. Three-dimensional numerical analysis of mixed ionic and electronic conducting cathode reconstructed by focused ion beam scanning electron microscope. *Journal of Power Sources*, 196(6):3073–3082, 2011.

[129] Z. Jiao and N. Shikazono. Simulation of solid oxide fuel cell anode microstructure evolution using phase field method. *Journal of The Electrochemical Society*, 160(6):F709–F715, 2013.

[130] T. Matsui, R. Kishida, J.-Y. Kim, H. Muroyama, and K. Eguchi. Performance deterioration of Ni–YSZ anode induced by electrochemically generated steam in solid oxide fuel cells. *Journal of The Electrochemical Society*, 157(5):B776–B781, 2010.

[131] T. Matsui, J.-Y. Kim, H. Muroyama, M. Shimazu, T. Abe, M. Miyao, and K. Eguchi. Anode microstructural change upon long-term operation for the cathode-supported tubular-type SOFC. *Solid State Ionics*, 225:50–54, 2012.

[132] M. Kubota, T. Okanishi, H. Muroyama, T. Matsui, and K. Eguchi. Microstructural evolution of ni–ysz cermet anode under thermal cycles with redox treatments. *Journal of The Electrochemical Society*, 162(4):F380–F386, 2015.

[133] M. Kishimoto, H. Iwai, M. Saito, and H. Yoshida. Quantitative evaluation of transport properties of sofc porous anode by random walk process. *ECS Transactions*, 25(2):1887–1896, 2009.

[134] D. Gostovic, J. R. Smith, D. P. Kundinger, K. S. Jones, and E.D. Wachsman. Three-dimensional reconstruction of porous LSCF cathodes. *Electrochemical and solid-state letters*, 10(12):B214–B217, 2007.

[135] J. R. Smith, A. Chen, D. Gostovic, D. Hickey, D. Kundinger, K. L. Duncan, R. T. DeHoff, K. S. Jones, and E. D. Wachsman. Evaluation of the relationship between cathode microstructure and electrochemical behavior for sofcs. *Solid State Ionics*, 180(1):90–98, 2009.

[136] P. R. Shearing, J. Gelb, and N. P. Brandon. X-ray nano computerised tomography of sofc electrodes using a focused ion beam sample-preparation technique. *Journal of the European Ceramic Society*, 30(8):1809–1814, 2010.

[137] Z. Chen, X. Wang, F. Giuliani, and A. Atkinson. Analyses of microstructural and elastic properties of porous SOFC cathodes based on focused ion beam tomography. *Journal of Power Sources*, 273:486–494, 2015.

[138] P. S. Jørgensen, K. V. Hansen, R. Larsen, and J. R. Bowen. A framework for automatic segmentation in three dimensions of microstructural tomography data. *Ultramicroscopy*, 110(3):216–228, 2010.

[139] P. S. Jørgensen, K. Yakal-Kremski, J. Wilson, J. R. Bowen, and S. Barnett. On the accuracy of triple phase boundary lengths calculated from tomographic image data. *Journal of Power Sources*, 261:198–205, 2014.

[140] L. Holzer, B. Münch, B. Iwanschitz, M. Cantoni, T. Hocker, and T. Graule. Quantitative relationships between composition, particle size, triple phase boundary length and surface area in nickel-cermet anodes for Solid Oxide Fuel Cells. *Journal of Power Sources*, 196(17):7076–7089, 2011.

[141] L. Holzer, B. Iwanschitz, T. Hocker, B. Münch, M. Prestat, D. Wiedenmann, U. Vogt, P. Holtappels, J. Sfeir, A. Mai, et al. Microstructure degradation of cermet anodes for solid oxide fuel cells: Quantification of nickel grain growth in dry and in humid atmospheres. *Journal of Power Sources*, 196(3):1279–1294, 2011.

[142] O. Pecho, L. Holzer, Z. Yáng, J. Martynczuk, T. Hocker, R. J. Flatt, and M. Prestat. Influence of strontium-rich pore-filling phase on the performance of $La_{0.6}Sr_{0.4}CoO_{3-\delta}$ thin-film cathodes. *Journal of Power Sources*, 274:295–303, 2015.

[143] G. Delette, J. Laurencin, F. Usseglio-Viretta, J. Villanova, P. Bleuet, E. Lay-Grindler, and T. Le Bihan. Thermo-elastic properties of SOFC/SOEC electrode materials determined from three-dimensional microstructural reconstructions. *International Journal of Hydrogen Energy*, 38(28):12379–12391, 2013.

[144] N. Vivet, S. Chupin, E. Estrade, T. Piquero, P. L. Pommier, D. Rochais, and E. Bruneton. 3d microstructural characterization of a solid oxide fuel cell anode reconstructed by focused ion beam tomography. *Journal of Power Sources*, 196(18):7541–7549, 2011.

[145] M. Cantoni and L. Holzer. Advances in 3D focused ion beam tomography. *Mrs Bulletin*, 39(04):354–360, 2014.

[146] P. S. Jørgensen and Jacob R. Bowen. Automatic quantitative image analysis of micrographs. *Proc. 8th Eur. Fuel Cell Forum*, page A0308, 2008.

[147] M. Kishimoto, H. Iwai, M. Saito, and H. Yoshida. Quantitative evaluation of solid oxide fuel cell porous anode microstructure based on focused ion beam and scanning electron microscope technique and prediction of anode overpotentials. *Journal of Power Sources*, 196(10):4555–4563, 2011.

[148] P. S. Jørgensen, K. V. Hansen, R. Larsen, and J. R. Bowen. High accuracy interface characterization of three phase material systems in three dimensions. *Journal of Power Sources*, 195(24):8168–8176, 2010.

[149] G. J. Nelson, W. M. Harris, J. J. Lombardo, J. R. Izzo, W. K. S. Chiu, P. Tanasini, M. Cantoni, C. Comninellis, J. C Andrews, Y. Liu, et al. Comparison of sofc cathode microstructure quantified using x-ray nanotomography and focused ion beam–scanning electron microscopy. *Electrochemistry Communications*, 13(6):586–589, 2011.

[150] R. Quey, H. Suhonen, J. Laurencin, P. Cloetens, and P. Bleuet. Direct comparison between x-ray nanotomography and scanning electron microscopy for the microstructure characterization of a solid oxide fuel cell anode. *Materials Characterization*, 78:87–95, 2013.

[151] D. Simwonis, F. Tietz, and D. Stöver. Nickel coarsening in annealed ni/8ysz anode substrates for solid oxide fuel cells. *Solid State Ionics*, 132(3):241–251, 2000.

[152] P. R. Shearing, Q. Cai, J. I. Golbert, V. Yufit, C. S. Adjiman, and N. P. Brandon. Microstructural analysis of a solid oxide fuel cell anode using focused ion beam techniques coupled with electrochemical simulation. *Journal of Power Sources*, 195(15):4804–4810, 2010.

[153] K. Yakal-Kremski, L. V. Mogni, A. Montenegro-Hernández, A. Caneiro, and S. A. Barnett. Determination of Electrode Oxygen Transport Kinetics Using Electrochemical Impedance Spectroscopy Combined with Three-Dimensional Microstructure Measurement: Application to $Nd_2NiO_{4+\delta}$. *Journal of The Electrochemical Society*, 161(14):F1366–F1374, 2014.

[154] J. R. Wilson, J. S. Cronin, and S. A. Barnett. Linking the microstructure, performance and durability of Ni-yttria-stabilized zirconia solid oxide fuel cell anodes using three-dimensional focused ion beam–scanning electron microscopy imaging. *Scripta Materialia*, 65(2):67–72, 2011.

[155] P. R. Shearing, J. Gelb, J. Yi, W.-K. Lee, M. Drakopolous, and N. P. Brandon. Analysis of triple phase contact in ni–ysz microstructures using non-destructive x-ray tomography with synchrotron radiation. *Electrochemistry Communications*, 12(8):1021–1024, 2010.

[156] R. Hill. Elastic properties of reinforced solids: some theoretical principles. *Journal of the Mechanics and Physics of Solids*, 11(5):357–372, 1963.

[157] M. Ostoja-Starzewski. Material spatial randomness: From statistical to representative volume element. *Probabilistic Engineering Mechanics*, 21(2):112–132, 2006.

[158] T. Kanit, S. Forest, I. Galliet, V. Mounoury, and D. Jeulin. Determination of the size of the representative volume element for random composites: statistical and numerical approach. *International Journal of solids and structures*, 40(13):3647–3679, 2003.

[159] P. Thevenaz, U. E. Ruttimann, and M. Unser. A pyramid approach to subpixel registration based on intensity. *Image Processing, IEEE Transactions on*, 7(1):27–41, 1998.

[160] MATLAB. *version 7.10.0 (R2010a)*. The MathWorks Inc., Natick, Massachusetts, United States, 2010.

[161] L. O'Gorman, M. J. Sammon, and M. Seul. *Practical algorithms for image analysis*. Cambridge University Press, 2008.

[162] W. Doyle. Operations useful for similarity-invariant pattern recognition. *Journal of the ACM (JACM)*, 9(2):259–267, 1962.

[163] C.A. Glasbey. "an analysis of histogram-based thresholding algorithms ". *CVGIP: Graphical Models and Image Processing*, 55(6):532 – 537, 1993.

[164] N. Otsu. A threshold selection method from gray-level histograms. *Automatica*, 11(285-296):23–27, 1975.

[165] J. P. Simmons, P. Chuang, M. Comer, J. E. Spowart, M. D. Uchic, and M. De Graef. Application and further development of advanced image processing algorithms for automated analysis of serial section image data. *Modelling and Simulation in Materials Science and Engineering*, 17(2):025002, 2009.

[166] I. Rotscholl. *Rekonstruktion einer zweiphasigen Elektrode mittels FIB/REM*. Bachelor thesis, Karlsruher Institut für Technologie (KIT), 2011.

[167] J. Canny. A computational approach to edge detection. *Pattern Analysis and Machine Intelligence, IEEE Transactions on*, 8(6):679–698, 1986.

[168] J. N. Kapur, P. K. Sahoo, and A. K. C. Wong. A new method for gray-level picture thresholding using the entropy of the histogram. *Vis. Graph. Image Process*, 29(3):273–285, 1985.

[169] H. Geisler, A. Kromp, A. Weber, and E. Ivers-Tiffée. Stationary FEM Model for Performance Evaluation of Planar Solid Oxide Fuel Cells Connected by Metal Interconnectors I. Model Framework and Validation. *Journal of The Electrochemical Society*, 161(6):F778–F788, 2014.

[170] J. Scott Cronin. *Three-Dimensional Structure Combined with Electrochemical Performance Analysis for Solid Oxide Fuel Cell Electrodes*. PhD thesis, Northwestern University, 2012.

[171] W. E. Lorensen and H. E. Cline. Marching cubes: A high resolution 3D surface construction algorithm. *Computer Graphics*, 21(4):163–169, 1987.

[172] L. Shen and Z. Chen. Critical review of the impact of tortuosity on diffusion. *Chemical Engineering Science*, 62(14):3748–3755, 2007.

[173] M. Ender, J. Joos, A. Weber, and E. Ivers-Tiffée. Anode microstructures from high-energy and high-power lithium-ion cylindrical cells obtained by X-ray nano-tomography. *Journal of Power Sources*, 269:912–919, 2014.

[174] P. R. Shearing, L. E. Howard, P. S. Jørgensen, N. P. Brandon, and S. J. Harris. Characterization of the 3-dimensional microstructure of a graphite negative electrode from a li-ion battery. *Electrochemistry communications*, 12(3):374–377, 2010.

[175] L. Holzer, O. Pecho, R. Flatt, M. Prestat, T. Hocker, G. Gaiselmann, M. Neumann, V. Schmidt, and B. Iwanschitz. Unraveling microstructure effects in Ni-YSZ anodes by 3D-analysis, FE-simulation and experimental characterization. *Proceedings of the 11th European SOFC and SOE Forum*, page B1104, 2014.

[176] W. Bangerth, R. Hartmann, and G. Kanschat. deal. II—a general-purpose object-oriented finite element library. *ACM Transactions on Mathematical Software (TOMS)*, 33(4):24, 2007.

[177] T. Carraro, J. Joos, B. Rüger, A. Weber, and E. Ivers-Tiffée. 3D finite element model for reconstructed mixed-conducting cathodes: II. Parameter sensitivity analysis. *Electrochimica Acta*, 77:309–314, 2012.

[178] C. C. Paige and M. A. Saunders. Solution of sparse indefinite systems of linear equations. *SIAM Journal on Numerical Analysis*, 12(4):617–629, 1975.

[179] R. Barrett, M. W. Berry, T. F. Chan, J. Demmel, J. Donato, J. Dongarra, V. Eijkhout, R. Pozo, C. Romine, and H. Van der Vorst. *Templates for the solution of linear systems: building blocks for iterative methods*, volume 43. Siam, 1994.

[180] M. Ender, J. Joos, T. Carraro, and E. Ivers-Tiffée. Quantitative characterization of $LiFePO_4$ cathodes reconstructed by FIB/SEM tomography. *Journal of The Electrochemical Society*, 159(7):A972–A980, 2012.

[181] L. Holzer, F. Indutnyi, P. H. Gasser, B. Münch, and M. Wegmann. Three-dimensional analysis of porous $BaTiO_3$ ceramics using FIB nanotomography. *Journal of Microscopy*, 216(1):84–95, 2004.

[182] J. Fleig. Solid oxide fuel cell cathodes: polarization mechanisms and modeling of the electrochemical performance. *Annual Review of Materials Research*, 33(1):361–382, 2003.

[183] C. Kreller, M. Drake, S. B. Adler, H.-Y. Chen, H.-C. Yu, K. Thornton, J. R. Wilson, and S. A. Barnett. Modeling sofc cathodes based on 3-d representations of electrode microstructure. *ECS Transactions*, 35(1):815–822, 2011.

[184] J. Jamnik and J. Maier. Generalised equivalent circuits for mass and charge transport: chemical capacitance and its implications. *Physical Chemistry Chemical Physics*, 3(9):1668–1678, 2001.

[185] J. Fleig and J. Maier. The polarization of mixed conducting SOFC cathodes: Effects of surface reaction coefficient, ionic conductivity and geometry. *Journal of the European Ceramic Society*, 24(6):1343–1347, 2004.

[186] Y. Lu, C. R. Kreller, S. B. Adler, J. R. Wilson, S. A. Barnett, P. W. Voorhees, H.-Y. Chen, and K. Thornton. Performance Variability and Degradation in Porous La1-xSrxCoO3-δ Electrodes. *Journal of The Electrochemical Society*, 161(4):F561–F568, 2014.

[187] L. C. R. Schneider, C. L. Martin, Y. Bultel, D. Bouvard, and E. Siebert. Discrete modelling of the electrochemical performance of sofc electrodes. *Electrochimica acta*, 52(1):314–324, 2006.

[188] L. C. R. Schneider, C. L. Martin, Y. Bultel, L. Dessemond, and D. Bouvard. Percolation effects in functionally graded sofc electrodes. *Electrochimica Acta*, 52(9):3190–3198, 2007.

[189] Q. Cai, C. S. Adjiman, and N. P. Brandon. Modelling the 3D microstructure and performance of solid oxide fuel cell electrodes: computational parameters. *Electrochimica Acta*, 56(16):5804–5814, 2011.

[190] Y. Suzue, N. Shikazono, and N. Kasagi. Micro modeling of solid oxide fuel cell anode based on stochastic reconstruction. *Journal of Power Sources*, 184(1):52–59, 2008.

[191] M. Kishimoto, H. Iwai, M. Saito, and H. Yoshida. Three-dimensional simulation of sofc anode polarization characteristics based on sub-grid scale modeling of microstructure. *Journal of The Electrochemical Society*, 159(3):B315–B323, 2012.

[192] M. E. Lynch, D. Ding, W. M. Harris, J. J Lombardo, G. J. Nelson, W. K. S. Chiu, and M. Liu. Flexible multiphysics simulation of porous electrodes: Conformal to 3D reconstructed microstructures. *Nano Energy*, 2(1):105–115, 2013.

[193] B. A. Boukamp and H. J. M. Bouwmeester. Interpretation of the Gerischer impedance in solid state ionics. *Solid State Ionics*, 157(1):29–33, 2003.

[194] B. Rüger, A. Weber, and E. Ivers-Tiffee. 3D-modelling and performance evaluation of mixed conducting (MIEC) cathodes. *ECS Transactions*, 7(1):2065–2074, 2007.

[195] COMSOL Multiphysics. *version 3.4*. COMSOL Inc, 2007.

[196] M. Søgaard, P. V. Hendriksen, T. Jacobsen, and M. B. Mogensen. Modelling of the polarization resistance from surface exchange and diffusion coefficient data. In *7th European Solid Oxide Fuel Cell Forum*, page B064, 2006.

[197] A. Häffelin. *Entwicklung eines 3D-FEM-Modells zur Untersuchung der Verlustmechanismen in mischleitenden Elektroden.* Diploma thesis, Karlsruher Institut für Technologie (KIT), 2012.

[198] R. Suwanwarangkul, E. Croiset, M. W. Fowler, P. L. Douglas, E. Entchev, and M. A. Douglas. Performance comparison of fick's, dusty-gas and stefan–maxwell models to predict the concentration overpotential of a sofc anode. *Journal of Power Sources*, 122(1):9–18, 2003.

[199] H. J. M. Bouwmeester, M. W. Den Otter, and B. A. Boukamp. Oxygen transport in La0. 6Sr0. 4Co1- y Fe y O3- δ. *Journal of Solid State Electrochemistry*, 8(9):599–605, 2004.

[200] J. Fleig. On the current–voltage characteristics of charge transfer reactions at mixed conducting electrodes on solid electrolytes. *Physical Chemistry Chemical Physics*, 7(9):2027–2037, 2005.

[201] A. Esquirol, N. P. Brandon, J. A. Kilner, and M. Mogensen. Electrochemical Characterization of $La_{0.6}Sr_{0.4}Co_{0.2}Fe_{0.8}O_3$ Cathodes for Intermediate-Temperature SOFCs. *Journal of The Electrochemical Society*, 151(11):A1847–A1855, 2004.

[202] D. Mantzavinos, A. Hartley, I. S. Metcalfe, and M. Sahibzada. Oxygen stoichiometries in La_{1-x} Sr_x Co_{1-y} $Fe_yO_{3-\delta}$ perovskites at reduced oxygen partial pressures. *Solid State Ionics*, 134(1):103–109, 2000.

[203] A. Schönbucher. *Thermische Verfahrenstechnik: Grundlagen und Berechnungsmethoden für Ausrüstungen und Prozesse; mit 103 Tabellen.* Springer-Verlag, 2002.

[204] J. Hayd and E. Ivers-Tiffée. Detailed electrochemical study on nanoscaled la0. 6sr0. 4coo3-δ sofc thin-film cathodes in dry, humid and co2-containing atmospheres. *Journal of The Electrochemical Society*, 160(11):F1197–F1206, 2013.

[205] Y. Saad. *Iterative methods for sparse linear systems.* Siam, 2003.

[206] A. Quarteroni and A. Valli. *Domain decomposition methods for partial differential equations.* Number CMCS-BOOK-2009-019. Oxford University Press, 1999.

[207] P. Ried, E. Bucher, W. Preis, W. Sitte, and P. Holtappels. Characterisation of $La_{0.6}Sr_{0.4}Co_{0.2}Fe_{0.8}O_{3-\delta}$ and $Ba_{0.5}Sr_{0.5}Co_{0.8}Fe_{0.2}O_{3-\delta}$ as Cathode Materials for the Application in Intermediate Temperature Fuel Cells. *ECS Transactions*, 7(1):1217–1224, 2007.

[208] A. Leonide, Y. Apel, and E. Ivers-Tiffee. SOFC modeling and parameter identification by means of impedance spectroscopy. *ECS Transactions*, 19(20):81–109, 2009.

[209] R. A. De Souza and J. A. Kilner. Oxygen transport in $La_{1-x}Sr_xMn_{1-y}Co_yO_{3\pm\delta}$ perovskites: Part I. Oxygen tracer diffusion. *Solid State Ionics*, 106(3):175–187, 1998.

[210] E. Girdauskaite, H. Ullmann, V. V. Vashook, U. Guth, G. B. Caraman, E. Bucher, and W. Sitte. Oxygen transport properties of$Ba_{0.5}Sr_{0.5}Co_{0.8}Fe_{0.2}O_{3-x}$ and $Ca_{0.5}Sr_{0.5}Mn_{0.8}Fe_{0.2}O_{3-x}$ obtained from permeation and conductivity relaxation experiments. *Solid State Ionics*, 179(11):385–392, 2008.

[211] E. Bucher, A. Egger, G. B. Caraman, and W. Sitte. Stability of the SOFC cathode material (Ba, Sr)(Co, Fe) $O_{3-\delta}$ in CO_2-containing atmospheres. *Journal of The Electrochemical Society*, 155(11):B1218–B1224, 2008.

[212] C. Niedrig, S. Taufall, M. Burriel, W. Menesklou, S. F. Wagner, S. Baumann, and E. Ivers-Tiffée. Thermal stability of the cubic phase in $Ba_{0.5}Sr_{0.5}Co_{0.8}Fe_{0.2}O_{3-\delta}$ (BSCF). *Solid State Ionics*, 197(1):25–31, 2011.

[213] V. Sonn, A. Leonide, and E. Ivers-Tiffée. Combined deconvolution and cnls fitting approach applied on the impedance response of technical ni/ 8ysz cermet electrodes. *Journal of The Electrochemical Society*, 155(7):B675–B679, 2008.

[214] M. W. Otter. *A study of oxygen transport in mixed conducting oxides using isotopic exchange and conductivity relaxation*. PhD thesis, Universiteit Twente, 2000.

[215] E. Bucher, G.B. Caraman, A. Egger, and W. Sitte. Oxygen Exchange Kinetics and Transport Properties of $La_{0.58}Sr_{0.4}Co_{0.2}Fe_{0.8}O_{3-\delta}$ for IT-SOFC Cathodes. *Proceedings of the 8th European Solid Oxide Fuel Cell Forum*, page A0406, 2008.

[216] C. Niedrig, S. F. Wagner, W. Menesklou, S. Baumann, and E. Ivers-Tiffée. Oxygen Equilibration Kinetics of Mixed-Conducting Perovskites BSCF, LSCF, and PSCF at 900°C Determined by Electrical Conductivity Relaxation. *Solid State Ionics*, page in press, 2015.

[217] A. Leonide, B. Rüger, A. Weber, W. A. Meulenberg, and E. Ivers-Tiffée. Impedance study of alternative (La, Sr)$FeO_{3-\delta}$ and (La, Sr)(Co, Fe)$O_{3-\delta}$ MIEC cathode compositions. *Journal of The Electrochemical Society*, 157(2):B234–B239, 2010.

[218] V. Wilde, H. Störmer, J. Szász, F. Wankmüller, E. Ivers-Tiffée, and D. Gerthsen. Effect of $Gd_{0.2}Ce_{0.8}O_2$ Sintering Temperature on Formation of a $SrZrO_3$ Blocking Layer between $Y_{0.16}Z_{0.84}O_2$, $Gd_{0.2}Ce_{0.8}O_2$ and $La_{0.58}Sr_{0.4}Co_{0.2}Fe_{0.8}O_{3-\delta}$. *ECS Transactions*, 66(2):103–107, 2015.

[219] J. Szász, F. Wankmüller, V. Wilde, H. Störmer, D. Gerthsen, N. H. Menzler, and E. Ivers-Tiffée. Nature and functionality of oxygen/cathode/electrolyte-interfaces in sofcs. *ECS Transactions*, 66(2):79–87, 2015.

[220] F. Wankmüller. *Fremdphasen an der Kathoden-/Elektrolyt-Grenzfläche von Feststoffelektrolyt Brennstoffzellen (SOFC)*. Master thesis, Karlsruher Institut für Technologie (KIT), 2015.

[221] A. Ali, X. Wen, K. Nandakumar, J. Luo, and K. T. Chuang. Geometrical modeling of microstructure of solid oxide fuel cell composite electrodes. *Journal of Power Sources*, 185(2):961–966, 2008.

[222] B. Kenney, M. Valdmanis, C. Baker, J. G. Pharoah, and K. Karan. Computation of tpb length, surface area and pore size from numerical reconstruction of composite solid oxide fuel cell electrodes. *Journal of Power Sources*, 189(2):1051–1059, 2009.

[223] C. Metcalfe, O. Kesler, T. Rivard, F. Gitzhofer, and N. Abatzoglou. Connected three-phase boundary length evaluation in modeled sintered composite solid oxide fuel cell electrodes. *Journal of the Electrochemical Society*, 157(9):B1326–B1335, 2010.

[224] F. Aurenhammer. Voronoi diagrams—a survey of a fundamental geometric data structure. *ACM Computing Surveys (CSUR)*, 23(3):345–405, 1991.

[225] H. Schaumburg. *Werkstoffe*. Springer-Verlag, 2013.

[226] S. Osher and R. P. Fedkiw. Level set methods: an overview and some recent results. *Journal of Computational physics*, 169(2):463–502, 2001.

[227] S. Osher and J. A. Sethian. Fronts propagating with curvature-dependent speed: algorithms based on Hamilton-Jacobi formulations. *Journal of computational physics*, 79(1):12–49, 1988.

[228] J. Bruchon, D. P. Muñoz, F. Valdivieso, S. Drapier, and G. Pacquaut. 3D simulation of the matter transport by surface diffusion within a Level-Set context. *European Journal of Computational Mechanics/Revue Européenne de Mécanique Numérique*, 19(1-3):281–292, 2010.

[229] J. R. Wilson, J. S. Cronin, A. T. Duong, S. Rukes, H.-Y. Chen, K. Thornton, D. R. Mumm, and S. Barnett. Effect of composition of ($La_{0.8}$ $Sr_{0.2}MnO_3$ –Y_2O_3-stabilized ZrO_2) cathodes: Correlating three-dimensional microstructure and polarization resistance. *Journal of Power Sources*, 195(7):1829–1840, 2010.

[230] S. B. Adler. Mechanism and kinetics of oxygen reduction on porous La_{1-x} Sr_{1-x} $CoO_{3-\delta}$ electrodes. *Solid State Ionics*, 111(1):125–134, 1998.

[231] Michael Becker. *Parameterstudie zur Langzeitbeständigkeit von Hochtemperaturbrennstoffzellen (SOFC)*. Mainz, 2007.

Werkstoffwissenschaft @ Elektrotechnik /
Universität Karlsruhe, Institut für Werkstoffe der Elektrotechnik

---

Band 1 Helge Schichlein
Experimentelle Modellbildung für die Hochtemperatur-Brennstoffzelle SOFC. 2003
ISBN 3-86130-229-2

Band 2 Dirk Herbstritt
Entwicklung und Optimierung einer leistungsfähigen Kathodenstruktur für die Hochtemperatur-Brennstoffzelle SOFC. 2003
ISBN 3-86130-230-6

Band 3 Frédéric Zimmermann
Steuerbare Mikrowellendielektrika aus ferroelektrischen Dickschichten. 2003
ISBN 3-86130-231-4

Band 4 Barbara Hippauf
Kinetik von selbsttragenden, offenporösen Sauerstoffsensoren auf der Basis von $Sr(Ti,Fe)O_3$. 2005
ISBN 3-86130-232-2

Band 5 Daniel Fouquet
Einsatz von Kohlenwasserstoffen in der Hochtemperatur-Brennstoffzelle SOFC. 2005
ISBN 3-86130-233-0

Band 6 Volker Fischer
Nanoskalige Nioboxidschichten für den Einsatz in hochkapazitiven Niob-Elektrolytkondensatoren. 2005
ISBN 3-86130-234-9

Band 7 Thomas Schneider
Strontiumtitanferrit-Abgassensoren. Stabilitätsgrenzen / Betriebsfelder. 2005
ISBN 3-86130-235-7

Band 8 Markus J. Heneka
Alterung der Festelektrolyt-Brennstoffzelle unter thermischen und elektrischen Lastwechseln. 2006
ISBN 3-86130-236-5

Die Bände sind im Verlagshaus Mainz (Aachen) erschienen.

Band 9 Thilo Hilpert
Elektrische Charakterisierung von Wärmedämmschichten mittels Impedanzspektroskopie. 2007
ISBN 3-86130-237-3

Band 10 Michael Becker
Parameterstudie zur Langzeitbeständigkeit von Hochtemperaturbrennstoffzellen (SOFC). 2007
ISBN 3-86130-239-X

Band 11 Jin Xu
Nonlinear Dielectric Thin Films for Tunable Microwave Applications. 2007
ISBN 3-86130-238-1

Band 12 Patrick König
Modellgestützte Analyse und Simulation von stationären Brennstoffzellensystemen. 2007
ISBN 3-86130-241-1

Band 13 Steffen Eccarius
Approaches to Passive Operation of a Direct Methanol Fuel Cell. 2007
ISBN 3-86130-242-X

Die Bände sind im Verlagshaus Mainz (Aachen) erschienen.

Fortführung als

Schriften des Instituts für Werkstoffe der Elektrotechnik, Karlsruher Institut für Technologie (ISSN 1868-1603)

bei KIT Scientific Publishing

Band 14 Stefan F. Wagner
Untersuchungen zur Kinetik des Sauerstoffaustauschs an modifizierten Perowskitgrenzflächen. 2009
ISBN 978-3-86644-362-4

Band 15 Christoph Peters
Grain-Size Effects in Nanoscaled Electrolyte and Cathode Thin Films for Solid Oxide Fuel Cells (SOFC). 2009
ISBN 978-3-86644-336-5

Band 16 Bernd Rüger
Mikrostrukturmodellierung von Elektroden für die Festelektrolytbrennstoffzelle. 2009
ISBN 978-3-86644-409-6

Band 17 Henrik Timmermann
Untersuchungen zum Einsatz von Reformat aus flüssigen Kohlenwasserstoffen in der Hochtemperaturbrennstoffzelle SOFC. 2010
ISBN 978-3-86644-478-2

Band 18 André Leonide
SOFC Modelling and Parameter Identification by Means of Impedance Spectroscopy. 2010
ISBN 978-3-86644-538-3

Band 19 Cornelia Endler-Schuck
Alterungsverhalten mischleitender LSCF Kathoden für Hochtemperatur-Festoxid-Brennstoffzellen (SOFCs). 2011
ISBN 978-3-86644-652-6

Band 20 Annika Utz
The Electrochemical Oxidation of H_2 and CO at Patterned Ni Anodes of SOFCs. 2011
ISBN 978-3-86644-686-1

Band 21 Jan Hayd
Nanoskalige Kathoden für den Einsatz in Festelektrolyt-Brennstoffzellen bei abgesenkten Betriebstemperaturen. 2012
ISBN 978-3-86644-838-4

Die Bände sind unter www.ksp.kit.edu als PDF frei verfügbar oder als Druckausgabe bestellbar.

Band 22 Michael Kornely
Elektrische Charakterisierung und Modellierung von metallischen Interkonnektoren (MIC) des SOFC-Stacks. 2012
ISBN 978-3-86644-833-9

Band 23 Dino Klotz
Characterization and Modeling of Electrochemical Energy Conversion Systems by Impedance Techniques. 2012
ISBN 978-3-86644-903-9

Band 24 Alexander Kromp
Model-based Interpretation of the Performance and Degradation of Reformate Fueled Solid Oxide Fuel Cells. 2013
ISBN 978-3-7315-0006-3

Band 25 Jan Philipp Schmidt
Verfahren zur Charakterisierung und Modellierung von Lithium-Ionen Zellen. 2013
ISBN 978-3-7315-0115-2

Band 26 Moses Ender
Mikrostrukturelle Charakterisierung, Modellentwicklung und Simulation poröser Elektroden für Lithiumionenzellen. 2014
ISBN 978-3-7315-0205-0

Band 27 Jörg Illig
Physically based Impedance Modelling of Lithium-Ion Cells. 2014
ISBN 978-3-7315-0246-3

Fortführung als

Schriften des Instituts für Angewandte Materialien – Werkstoffe der Elektrotechnik Karlsruher Institut für Technologie (ISSN 2365-8029)

bei KIT Scientific Publishing

Band 28 Christian Niedrig
Electrochemical Performance and Stability of $Ba_{0.5}Sr_{0.5}Co_{0.8}Fe_{0.2}O_{3-\delta}$ for Oxygen Transport Membranes. 2015
ISBN 978-3-7315-0437-5

Die Bände sind unter www.ksp.kit.edu als PDF frei verfügbar oder als Druckausgabe bestellbar.

Band 29 Daniel Manka
Elektrochemisch-optische Impedanzspektroskopie. 2016
ISBN 978-3-7315-0547-1

Band 30 Jochen Joos
Microstructural Characterisation, Modelling and Simulation of Solid Oxide Fuel Cell Cathodes. 2017
ISBN 978-3-7315-0625-6